Lecture Notes in Mathematics

Edited by A. Dold and B. Eckmann

495

Adalbert Kerber

T0222480

Representations of Permutation Groups II

Springer-Verlag
Berlin Heidelberg New York 1975

Autor
Prof. Dr. Adalbert Kerber
Lehrstuhl D für Mathematik
Rhein.-Westf. Technische
Hochschule Aachen
Templergraben 55
51 Aachen/BRD

Library of Congress Cataloging in Publication Data

Kerber, Adalbert.
Representations of permutation groups I-II.

(Lecture notes in mathematics, 240, 495)
Bibliography: p.
Includes indexes.
CONTENTS: pt. 1. Representation of wreath products and applications to the representation theory of symmetric and alternating groups.
1. Permutation groups. 2. Representations of groups. I. Title. II. Series: Lecture notes in mathematics (Berlin), 240, etc.
QA3.L28 no. 240, etc. 510'.8s [512'.2] 72-183956

AMS Subject Classifications (1970): 05A15, 20C30

ISBN 3-540-07535-6 Springer-Verlag Berlin · Heidelberg · New York
ISBN 0-387-07535-6 Springer-Verlag New York · Heidelberg · Berlin

Printed in Germany
Offsetdruck: Julius Beltz, Hemsbach/Bergstr.

Preface

The description of the representation theory of wreath products and its applications are continued in this second part.

In part I the emphasis lay on the construction of the irreducible matrix representations of wreath products over an algebraically closed field. In part II, I consider mainly the ordinary irreducible characters of these groups, which were less important in part I.

The considerations apply especially to the symmetrization of representations, so that we obtain quite easily famous results of Schur, Frobenius, Weyl and van der Waerden about the connection between the representation theories of general linear and symmetric groups. They apply also to the theory of enumeration under group action so that we obtain the most important results of this theory, which has been developed mainly by Redfield, Pôlya and de Bruijn. This theory is nowadays an essential part of combinatorics and it yields the main enumeration techniques in graph theory.

These applications and some related topics are discussed here.

In the first sections the main results needed from part I are quoted, so that this part is in a sense also selfcontained.

I would like to express my sincerest thanks to many colleagues who work in that field and in particular to the people working at the "Lehrstuhl D für Mathematik der RWTH Aachen" for very helpful and stimulating discussions and cooperation.

Adalbert Kerber

Contents

Introduction

Having described in part I a construction of the irreducible representations of the wreath product $G\sim H$ (G a finite group, H a permutation group of finite degree) over an algebraically closed field, we now consider their characters.

The character formulae obtained have various applications. They yield directly the characters of several permutation representations of $G\sim H$ (if G is also a permuation group) which are of interest in combinatorics, e.g. the composition $H[G]$, the exponentiation $[G]^H$ and the matrix group $[H;G]$. Having obtained the permutation character, it is in each case not difficult to obtain the cycle-index. Furthermore the way how the character formulae are derived gives a hint how to obtain enumeration theorems in weighted form for the groups in question.

The applications to representation theory are based on the fact that $G\times H$ has a nice embedding in $G\sim H$: $G\times H \simeq \mathrm{diag}G^*\cdot H'$, where G^* denotes the basis group of $G\sim H$, H' a certain complement which is isomorphic to H, and $\mathrm{diag}G^*$ the diagonal of the basis group. Hence a representation F of G yields a representation $\overset{\widetilde{n}}{\#}F$ of $G\sim H$ (if $H \leq S_n$), as it is described in part I, and the restrictions

$$\hat{F} := \overset{\widetilde{n}}{\#}F \downarrow \mathrm{diag}G^* \quad \text{and} \quad \check{F} := \overset{\widetilde{n}}{\#}F \downarrow H'$$

of this representation yield representations of G and H, the elements of which are mutually centralizing.

Hence Schur's lemma (if the groundfield is so that it can be applied) shows a close connection between the decompositions of $\hat{F}$ and $\check{F}$. It shows how $\hat{F}$, which is in fact equal to $\overset{n}{\otimes} F$, if $H \leq S_n$, can be "symmetrized" with each irreducible representation, which is an irreducible constituent of $\check{F}$.

Doing this for $G := GL(m,C)$, $F := id_{GL(m,C)}$, $H := S_n$, we obtain famous results about the close connection between the representation theories of symmetric and general linear groups.

There are also further and sometimes quite surprising applications.

Chapter I

Characters of Wreath Products

In the first section of this chapter the definition of the wreath product $G\sim H$ of a group G and a permutation group H of finite degree is recalled from part I as well as the results about conjugacy classes of wreath products of the form $G\sim S_n$.

To these results about conjugacy classes some recent results concerning their splitting over certain normal subgroups of index 2 are added.

The second section contains the basic results of part I concerning the construction of matrix representations of wreath products as well as some character formulae which are applied in the following chapters.

For an example, the ordinary irreducible representations of hyperoctahedral groups $S_2\sim S_n$ as well as their splitting over certain normal subgroups of index 2 is described in detail. This corresponds to the results of the first section which concern the splitting of conjugacy classes and it covers the ordinary representation theory of the series of Weyl groups. The results on the splitting of ordinary irreducible representations is then applied to the evaluation of decomposition matrices.

1. Repetition and more about conjugacy classes of certain wreath products

If G is a group, H a permutation group acting on the set of symbols $\mathbb{N}_n := \{1,\ldots,n\}$, we consider the set $G\sim H$ of all ordered pairs $(f;\pi)$ consisting of a map f from $\mathbb{N}_n$ to G (for short: $f : \mathbb{N}_n \to G$ or $f \in G^{\mathbb{N}_n}$) and a permutation $\pi \in H$:

$$G\sim H := G^{\mathbb{N}_n} \times H = \{(f;\pi) \mid f : \mathbb{N}_n \to G \wedge \pi \in H\}.$$

For such maps $f, f' : \mathbb{N}_n \to G$ and elements $\pi \in H$ we define $f^{-1}, f_\pi, ff', e : \mathbb{N}_n \to G$ by

$$f^{-1} : \mathbb{N}_n \to G : i \mapsto f(i)^{-1},$$

$$f_\pi : \mathbb{N}_n \to G : i \mapsto f(\pi^{-1}(i)),$$

$$ff' : \mathbb{N}_n \to G : i \mapsto f(i)f'(i),$$

$$e : \mathbb{N}_n \to G : i \mapsto 1_G.$$

Since we agreed upon reading products of permutations from the right to the left:

$$\pi\pi' : \mathbb{N}_n \to \mathbb{N}_n : i \mapsto \pi(\pi'(i)),$$

we obtained

$$\forall\ f,\pi,\pi' \ ((f_\pi)_{\pi'} = f_{\pi'\pi}).$$

Using this result, it is easy to check, that $G\sim H$ together with the composition defined by

$$\forall\ f,f',\pi,\pi' \ ((f;\pi)(f';\pi') := (ff'_\pi;\pi\pi'))$$

is a group, <u>the wreath product of</u> G <u>with</u> H.

Since

$$\forall\ f,\pi\ ((f_{\pi^{-1}})^{-1} = (f^{-1})_{\pi^{-1}} =: f^{-1}_{\pi^{-1}}),$$

we obtained for the identity element $1_{G\wr H}$ of $G\wr H$ and the inverse of $(f;\pi)$:

$$1_{G\wr H} = (e;1_H) \wedge (f;\pi)^{-1} = (f^{-1}_{\pi^{-1}};\pi^{-1}).$$

The order of this group is

$$|G\wr H| = |G|^n|H|.$$

Let us now mention some interesting subgroups of $G\wr H$.

The normal subgroup

$$G^* := \{(f;1_H)\,|\,f : \mathbb{N}_n \to G\} \trianglelefteq G\wr H$$

was called the <u>basis group</u> of $G\wr H$. It is the inner direct product of n subgroups G_i which are isomorphic to G:

$$G \simeq G_i := \{(f;1_H)\,|\ \forall\ j \neq i\ (f(j) = 1_G)\},\ 1 \leq i \leq n,$$

$$G^* = G_1 \times \ldots \times G_n = \mathop{\times}_{i=1}^{n} G_i .$$

The subgroup

$$H' := \{(e;\pi)\ |\ \pi \in H\} \simeq H,$$

which is isomorphic to H is a complement of G^*:

$$G\wr H = G^*H' \wedge G^* \trianglelefteq G\wr H \wedge G^* \cap H' = \{1_{G\wr H}\}.$$

It is sometimes useful to describe $(f;\pi)$ more explicitly by displaying the values of f, i.e. to write

$$(f(1),\dots,f(n);\pi)$$

instead of $(f;\pi)$. Using this notation, the <u>diagonal of the basis group</u> is:

$$\begin{aligned}\mathrm{diag}G^* &:= \{(f;1_H) \mid f \text{ constant}\}\\ &= \{(g,\dots,g;1_H) \mid g \in G\} \simeq G.\end{aligned}$$

Multiplying this by H' we obtain a further subgroup of interest:

$$\begin{aligned}(\mathrm{diag}G^*)H' &= \{(f;\pi) \mid f \text{ constant}\}\\ &= \{(g,\dots,g;\pi) \mid g \in G \wedge \pi \in H\} \simeq G \times H.\end{aligned}$$

If $C(G)$ denotes the centre of G, we obtain for the centre $C(G\wr H)$ of $G\wr H$:

1.1 $G \neq \{1_G\} \Rightarrow$

$C(G\wr H) = \{(f;1_H) \mid f{:}N_n \to C(G) \wedge f \text{ constant on each orbit of } H\}.$

More special subgroups of $G\wr H$ can be defined using given normal subgroups M of G, N of H, which are of index ≤ 2 in G and H, respectively:

1.2 Def.: $M \trianglelefteq G$, $N \trianglelefteq H$ and $|G:M| \leq 2 \geq |H:N|$, then

(i) $G\wr H_M := \{(f;\pi) \mid \prod_1^n f(i) \in M\} \trianglelefteq G\wr H,$

(ii) $G \wr H_M^N := \{(f;\pi) \mid \prod_1^n f(i) \in \begin{cases} M, \text{ if } \pi \in N \\ G \setminus M, \text{ if } \pi \in H \setminus N \end{cases}\} \trianglelefteq G \wr H.$

It is easy to check, that the following is valid:

1.3 If M, N are subgroups of index ≤ 2 in G, H, respectively, then $G \wr N$, $G \wr H_M$ and $G \wr H_M^N$ are subgroups of index ≤ 2 in $G \wr H$.

Special cases of such subgroups have been considered in part I section 6. To see this we use the following faithful permutation representation φ of $G \wr H$, if $G \leq S_m$ and $H \leq S_n$:

1.4 $\varphi : G \wr H \to S_{mn} : (f;\pi) \mapsto \begin{pmatrix} (j-1)m + i \\ (\pi(j)-1)m + f(\pi(j))(i) \end{pmatrix}_{\substack{1 \leq i \leq m. \\ 1 \leq j \leq n}}$

Then if we denote by P^+ the subgroup of even permutations of a given permutation group P, the following holds:

1.5 $G \leq S_m$, then

(i) if m is even, we have:

$\varphi[G \wr H_{G^+}] = \varphi[G \wr H] \cap A_{mn} = \varphi[G \wr H]^+,$

(ii) and if m is odd, then

$\varphi[G \wr H_{G^+}^{H^+}] = \varphi[G \wr H] \cap A_{mn} = \varphi[G \wr H]^+.$

Hence e.g.

$$2|m \Rightarrow \varphi[S_m \wr S_{n_{A_m}}] = \varphi[S_m \wr S_n]^+,$$

$$2\nmid m \Rightarrow \varphi[S_m \wr S_{n_{A_m}}^{A_n}] = \varphi[S_m \wr S_n]^+.$$

The subgroup $\varphi[S_m \wr S_n]^+$ was denoted by $S_m \wr S_n^{\,+}$ in section 6 of part I.

It should be mentioned, that the series of Weyl groups but not the exceptional Weyl groups E_6, E_7, E_8, F_4 and G_2 are special cases of the groups defined above. The Weyl groups are defined to be certain permutation groups acting on certain subsets ("root systems") of euclidean spaces. It can be shown, that the following is valid (cf. e.g. Humphreys [1], 12.1, Stewart [1], chapter 7):

1.6 (i) The Weyl group of type A_n $(n \geq 1)$ is similar (cf. I, p.29) to S_{n+1}.

(ii) The Weyl group of type B_n $(n \geq 2)$ is similar to $\varphi[S_2 \wr S_n]$ as is the Weyl group of type C_n $(n \geq 3)$.

(iii) The Weyl group of type D_n $(n \geq 4)$ is similar to $\varphi[S_2 \wr S_n] \cap A_{2n} = \varphi[S_2 \wr S_n]^+ = \varphi[S_2 \wr S_{n_{A_2}}]$.

Hence the following considerations can be applied especially to the series of Weyl groups.

Having mentioned certain subgroups of $G\sim H$, let us now briefly recall some notation which was introduced when conjugacy classes of wreath products of the form $G\sim S_n$ were considered.

We agreed on writing permutations $\pi \in S_n$ as products of disjoint cycles π_ν as usual:

$$\underline{\underline{1.7}} \qquad \pi = \prod_{\nu=1}^{c(\pi)} \pi_\nu = \prod_1^{c(\pi)} (j_\nu \ \pi(j_\nu) \ldots \pi^{k_\nu - 1}(j_\nu)),$$

where in each of these cyclic factors $\pi_\nu = (j_\nu \ldots \pi^{k_\nu-1}(j_\nu))$ j_ν is the least symbol which is included and where the numbering of the cycles is uniquely determined by assuming

$$j_1 \leq j_2 \leq \cdots \leq j_{c(\pi)}.$$

With these conventions, 1.7 is uniquely determined. If now $(f;\pi) \in G\sim H$, so that $f : \mathbb{N}_n \to G$, then the element

$$\underline{\underline{1.8}} \quad g_\nu(f;\pi) := ff_\pi \ldots f_{\pi^{k_\nu-1}}(j_\nu) = f(j_\nu)f(\pi^{-1}(j_\nu)) \ldots f(\pi^{-k_\nu+1}(j_\nu)) \in G$$

was called <u>the cycle product associated with the ν-th cyclic factor</u> π_ν <u>of</u> π <u>with respect to</u> f. We assume now, that G has countably many conjugacy classes and that $C^1, C^2, \ldots$ is a fixed numbering of them.

If in $(f;\pi) \in G\sim H$ the permutation π is of type

$$T\pi = (a_1(\pi), \ldots, a_n(\pi)),$$

i.e. if $a_k(\pi)$ is the number of cyclic factors π_ν of π

which are of length $k_\nu = k$, $1 \leq k \leq n$, let

<u>1.9</u> $\pi^j_{ik} = (r^j_{ik}\ \pi(r^j_{ik}) \ldots \pi^{k-1}(r^j_{ik}))$, $1 \leq j \leq a_{ik}(f;\pi)$,

be the cyclic factors of length k whose associated cycle products

<u>1.10</u> $$g^j_{ik}(f;\pi) := f \ldots f_{\pi^{k-1}}(r^j_{ik})$$

belong to C^i, when occasion arises.

The notation 1.9 is uniquely determined, if we again agree that r^j_{ik} is the least of the occuring symbols and that if $a_{ik}(f;\pi) \geq 2$ we have

$$\forall\ 1 \leq j < a_{ik}(f;\pi)\ (r^j_{ik} < r^{j+1}_{ik}).$$

Then

$$\pi = \prod_{\substack{i,j,k \\ a_{ik}(f;\pi)>0}} \pi^j_{ik}.$$

The scheme

<u>1.11</u> $T(f;\pi) := (a_{ik}(f;\pi))$, $\quad$ i row index, k column index, $1 \leq k \leq n$,

was called <u>the type of</u> $(f;\pi)$. Conjugate elements are of the same type (cf. I 3.7):

<u>1.12</u> <u>$\forall\ (f;\pi),(f';\pi') \in G\wr H\ ((f;\pi) \sim (f';\pi') \Rightarrow T(f;\pi) = T(f';\pi'))$.</u>

In case $H = S_n$, the converse holds as well ,so that two elements of $G\sim S_n$ are conjugates if and only if they are of the same type:

$$\underline{\underline{1.13}} \quad \underline{\forall (f;\pi),(f';\pi') \in G\sim S_n \ ((f;\pi) \sim (f';\pi') \Leftrightarrow T(f;\pi) = T(f';\pi')).}$$

In addition to these results which have been proved in part I let us now answer the question, which of these conjugacy classes of $G\sim S_n$ split into several conjugacy classes of $G\sim A_n$, of $G\sim S_{n_M}$, and of $G\sim S_{n_M}^{A_n}$, respectively, where M denotes a given subgroup of index 2 in G.

Since these three subgroups are of index ≤ 2 in $G\sim S_n$, we know that such splitting conjugacy classes split into two conjugacy classes of equal order, and that this happens if and only if the subgroup U of $G\sim S_n$ considered is of index 2 and the centralizers of an element $(f;\pi)$ of that conjugacy class $C^{G\sim S_n}(f;\pi)$ in $G\sim S_n$ and U are equal:

$$C_{G\sim S_n}((f;\pi)) = C_U((f;\pi)).$$

Hence a consideration of the centralizers of elements in $G\sim S_n$ may provide an answer to the question. Let us briefly recall, what has been said in part I about the centralizer of an element in $G\sim S_n$.

Centralizers of conjugate elements are conjugate subgroups, hence 1.13 implies that to describe the centralizer of any element of type $(a_{ik}(f;\pi))$, we may assume without restriction, that the map f has value 1_G everywhere except possibly at the points r^j_{ik} (cf. I 3.22). Thus we may assume, that for each cyclic factor $\pi^j_{ik} = (r^j_{ik} \ldots \pi^{k-1}(r^j_{ik}))$ of π (cf. 1.9) we have (if $k \geq 2$):

$$f(r^j_{ik}) \in C^i \wedge f(\pi^{-1}(r^j_{ik})) = \ldots = f(\pi^{-k+1}(r^j_{ik})) = 1_G,$$

so that if $a_{ik}(f;\pi) > 0$:

<u>1.14</u> $\forall\, i,k,\ 1 \leq j \leq a_{ik}(f;\pi)\ (f(r^j_{ik}) = g^j_{ik}(f;\pi))$.

We define corresponding maps $f^j_{ik} : \mathbb{N}_n \to G$ by

$$f^j_{ik}(s) := \begin{cases} f(r^j_{ik}), & \text{if } s = r^j_{ik} \\ 1_G, & \text{elsewhere.} \end{cases}$$

Then the following is obviously valid:

<u>1.15</u> (i) $(f;\pi) = \prod\limits_{\substack{i,j,k \\ a_{ik}(f;\pi)>0}} (f^j_{ik};\pi^j_{ik})$, and

(ii) <u>the factors $(f^j_{ik};\pi^j_{ik})$ of $(f;\pi)$ are pairwise commutative.</u>

For the factor $(f^j_{ik};\pi^j_{ik})$ we have

$$(f^j_{ik};\pi^j_{ik}) \in \underbrace{(G_{r^j_{ik}} \times G_{\pi(r^j_{ik})} \times \ldots \times G_{\pi^{k-1}(r^j_{ik})})S'_k}_{\simeq\, G\wr S_k} \leq G\wr S_n.$$

Let us denote the corresponding element of $G\sim S_k$ by

$$(f_j^{ik};\pi_j^{ik}).$$

Then 1.15 yields for the centralizer of $(f;\pi)$:

1.16 The centralizer $C_{G\sim S_n}(f;\pi)$ of $(f;\pi)$ in $G\sim S_n$ is a subgroup of $G\sim S_n$ which is an embedding of

$$\underset{\substack{i,k\\ a_{ik}(f;\pi)>0}}{\times} C_{G\sim S_k}(f_1^{ik};\pi_1^{ik})\sim S_{a_{ik}(f;\pi)}$$

To decide, whether such a subgroup of $G\sim S_n$ is contained in a given normal subgroup U, we need a closer examination of its direct factors.

The centralizer of

$$(f_1^{ik};\pi_1^{ik}) = (\underbrace{g_{ik}^1(f;\pi),\ 1_G,\dots,1_G}_{\text{the } k \text{ values of } f_1^{ik}};\pi_1^{ik}\ )$$

was described explicitly in I 3.19:

1.17 $C_{G\sim S_k}(f_1^{ik};\pi_1^{ik}) = \mathrm{diag}(C_G(g_{ik}^1(f;\pi))*)\langle(f_1^{ik};\pi_1^{ik})\rangle.$

Let us consider the elements of this group. We consider first the special case

$$(f_1^{ik};\pi_1^{ik}) := (g,1,\dots,1;(1\dots k)).$$

Then the elements of $\mathrm{diag}(C_G(g)^*)\langle(g,1,\ldots,1;(1\ldots k))\rangle$ are of the form

$$(c,\ldots,c;1)(g,1,\ldots,1;(1\ldots k))^l = (\underbrace{cg,\ldots,cg}_{l\text{-times}},c,\ldots c;(1\ldots k)^l),$$

where c denotes an element of $C_G(g)$.

Hence the elements of $C_{G\sim S_k}(f_1^{ik};\pi_1^{ik})\sim S_{a_{ik}(f;\pi)}$ are conjugate to elements of the form

$$x:=(\ldots,(c_jg,\ldots,c_jg,c_j,\ldots,c_j;(1\ldots k)^{l_j}),\ldots;\rho)\in(G\sim S_k)\sim S_{a_{ik}(f;\pi)},$$

where $\rho \in S_{a_{ik}(f;\pi)}$ and $g = f(r_{ik}^1)$ and $c_j \in C_G(g)$, $1 \leq j \leq a_{ik}(f;\pi)$.

If we apply the associativity of the wreath product (cf. the proof of 2.29 in I), then x corresponds to the element

$$y:=(\ldots,c_jg,\ldots,c_jg,c_j,\ldots,c_j,\ldots;((1\ldots k)^{l_1},\ldots,(1\ldots k)^{l_{a_{ik}(f;\pi)}};\rho)$$

in $G\sim S_{ka_{ik}(f;\pi)}$.

This altogether yields the crucial lemma:

1.18 <u>The centralizer of $(f;\pi) \in U \trianglelefteq G\sim S_n$ in $G\sim S_n$ is contained in the normal subgroup U of $G\sim S_n$ if and only if for all the pairs (i,k) with $a_{ik}(f;\pi)>0$ we have that</u>

all the elements

$$(\dots,c_jf(r^1_{ik}),\dots,c_jf(r^1_{ik}),c_j,\dots;((1\dots k)^{l_1},\dots,(1\dots k)^{l_{a_{ik}(f;\pi)}};\rho)),$$

where $c_j \in C_G(f(r^1_{ik})), 1 \leq j \leq a_{ik}(f;\pi)$, and $\rho \in S_{a_{ik}(f;\pi)}$,

are contained in U.

In order to complete the proof of 1.18, we need only notice that the normality of U implies, that 1.14 can be assumed without restriction.

Let us apply this to several special cases.

1.19 The conjugacy class of $(f;\pi) \in G\sim A_n \trianglelefteq G\sim S_n$ in $G\sim S_n$ splits over $G\sim A_n$ if and only if:

(i) $2 \nmid k \Rightarrow a_{ik}(f;\pi) \leq 1$, and

(ii) $2 \mid k \Rightarrow a_{ik}(f;\pi) = 0$.

Proof: The element

1.20 $$(\dots,c_jf(r^1_{ik}),\dots,c_jf(r^1_{ik}),c_j,\dots;(\dots,(1\dots k)^{l_j},\dots;\rho))$$

is contained in $G\sim A_n$ if and only if the element

1.21 $$\varphi((1\dots k)^{l_1},\dots,(1\dots k)^{l_{a_{ik}(f;\pi)}};\rho)) \in S_n$$

is contained in A_n, where φ denotes the permutation representation 1.4 applied to $S_k \sim S_{a_{ik}(f;\pi)}$.

If k is odd, then 1.21 is always contained in A_n if and only if $a_{ik}(f;\pi) \leq 1$, for otherwise the choice $l_1 = \ldots = l_{a_{ik}(f;\pi)} := 0$ and $\rho := (12)$ would yield the element

$$\varphi(e;(12)) = (1,k+1)(2,k+2)\ldots(k,2k) \in S_n \backslash A_n,$$

which is of form 1.21.

If k is even, then 1.21 is always contained in A_n if and only if $a_{ik}(f;\pi) = 0$, since otherwise the choice $l_1 := 1, l_2 = \ldots = l_{a_{ik}(f;\pi)} := 0$, $\rho := 1$, would yield the element

$$\varphi((1\ldots k),1,\ldots,1;1) = (1\ldots k) \in S_n \backslash A_n,$$

which is of form 1.21.

q.e.d.

1.22 <u>If M denotes a subgroup of index 2 in G, then the conjugacy class of $(f;\pi) \in G \sim S_{n_M} \trianglelefteq G \sim S_n$ in $G \sim S_n$ splits over $G \sim S_{n_M}$ if and only if:</u>

(i) <u>$2 | k \wedge a_{ik}(f;\pi) > 0 \Rightarrow C^i \subseteq M$, and</u>

(ii) <u>$2 \nmid k \wedge a_{ik}(f;\pi) > 0 \Rightarrow C^i$ splits over M.</u>

<u>Proof:</u> The element 1.20 is contained in $G\wr S_{n_M}$ if and only if the element

1.23 $$(\ldots,\underbrace{\underbrace{c_j f(r^1_{ik}),\ldots,c_j f(r^1_{ik})}_{l_j\text{-times}},c_j,\ldots,c_j}_{k \text{ values}},\ldots;(e;1))$$

is contained in $G\wr S_{n_M}$, and this happens if and only if the element

1.24 $$\prod_j (c_j)^k f(r^1_{ik})^{l_j}$$

is contained in M.

This must be satisfied for given $f(r^1_{ik})$ and each choice of $c_j \in C_G(f(r^1_{ik}))$ and of exponents k and l_j.

Hence we need $f(r^1_{ik}) \in M$ since one might choose $l_j = 1$ and $c_j := 1$ in 1.24.

If $f(r^1_{ik}) \in M$ then we still need, that the elements

1.25 $$\prod_j (c_j)^k$$

are contained in M.

In case of odd k, this needs $c_j \in M$, i.e. it needs $C_G(f(r^1_{ik})) \subseteq M$, i.e. it needs the splitting of the conjugacy class C^i of $f(r^1_{ik})$ over M.

If on the other hand $f(r^1_{ik}) \in M$ and if C^i splits in case k is odd, then 1.23 is contained in $G\wr S_{n_M}$.

q.e.d.

<u>1.26 If M denotes a subgroup of index 2 in G, then the conjugacy class of $(f;\pi) \in G\wr S_{n_M}^{A_n} \trianglelefteq G\wr S_n$ in $G\wr S_n$ splits over $G\wr S_{n_M}^{A_n}$ if and only if:</u>

(i) $2|k \wedge a_{ik}(f;\pi) > 0 \Rightarrow C^i \subseteq G\backslash M$, and

(ii) $2\nmid k \wedge a_{ik}(f;\pi) > 0 \Rightarrow a_{ik}(f;\pi) = 1 \wedge C^i$ splits over M.

Proof: If 1.20 is always contained in $G\wr S_{n_M}^{A_n}$, then $\varphi(e;\rho)$ must be even for all $\rho \in S_{a_{ik}(f;\pi)}$, i.e.

$$2\nmid k \wedge a_{ik}(f;\pi) > 0 \Rightarrow a_{ik}(f;\pi) = 1.$$

If this is satisfied, then for $2\nmid k$ and $a_{ik}(f;\pi) = 1$ we have

$$\varphi((1...k)^{l_1};1) \in A_k, \quad \forall\, l_1.$$

Hence we have furthermore that

$$(c_1)^k f(r_{ik}^1) \in M.$$

This needs for odd k and $a_{ik}(f;\pi) = 1$:

$$f(r_{ik}^1) \in M \wedge c_1 \in M.$$

This altogether yields the necessary condition for the

splitting of the conjugacy class of $(f;\pi)$:

$$2\nmid k \wedge a_{ik}(f;\pi) > 0 \Rightarrow a_{ik}(f;\pi) = 1 \wedge C^i \text{ splits over } M.$$

I.e.: (ii) is a necessary condition.

If on the other hand we have for an even k: $a_{ik}(f;\pi) > 0$, then we need (put $l_1 := 1$, $l_2 = \ldots = l_{a_{ik}(f;\pi)} := 0$, $c_1 = \ldots = c_{a_{ik}(f;\pi)} := 1$) that $f(r_{ik}^1) \in G\backslash M$, i.e. we obtain

$$2 \mid k \wedge a_{ik}(f;\pi) > 0 \Rightarrow C^i \subseteq G\backslash M,$$

i.e. (i) is necessary as well.

It is obvious that, if (i) and (ii) are satisfied, then all the elements 1.20 are contained in $G\sim S_{n_M}^{A_n}$.

q.e.d.

A special case is formed by the hyperoctahedral groups: $G := S_2$. The types of elements of $S_2 \sim S_n$ are matrices with 2 rows and n columns. Let their first row belong to the conjugacy class $C^1 := \{1_{S_2}\}$ and their second one to $C^2 := \{(12)\}$.

Then we obtain the following corollary from 1.19, 1.22 and 1.26:

1.27 <u>The conjugacy class of $(f;\pi) \in S_2 \sim S_n$ of type $(a_{ik}(f;\pi))$ splits</u>

(i) <u>over $S_2 \sim A_n$ if and only if $(a_{ik}(f;\pi))$ is of the form</u>

$$\begin{pmatrix} (\leq 1) & 0 & (\leq 1) & 0 & \ldots \\ (\leq 1) & 0 & (\leq 1) & 0 & \ldots \end{pmatrix},$$

(ii) <u>over $S_{2} \sim S_{n_{A_2}}$ if and only if $(a_{ik}(f;\pi))$ is of the form</u>

$$\begin{pmatrix} 0 & * & 0 & * & \ldots \\ 0 & 0 & 0 & 0 & \ldots \end{pmatrix},$$

(iii) <u>over $S_2 \sim S_{n_{A_2}}^{A_n}$ if and only if $(a_{ik}(f;\pi))$ is of the form</u>

$$\begin{pmatrix} 0 & 0 & 0 & 0 & \ldots \\ 0 & * & 0 & * & \ldots \end{pmatrix}.$$

This corollary includes the results of Young on the conjugacy class of $S_2 \sim S_{n_{A_2}}$ (Young [1], cf. also Carter [1], [2], Mayer [1], Taylor [1], [2]).

For a numerical example we consider $S_2 \sim S_4$. Its types are

$$\begin{pmatrix}4&0&0&0\\0&0&0&0\end{pmatrix}, \begin{pmatrix}3&0&0&0\\1&0&0&0\end{pmatrix}, \begin{pmatrix}2&0&0&0\\2&0&0&0\end{pmatrix}, \begin{pmatrix}1&0&0&0\\3&0&0&0\end{pmatrix}, \begin{pmatrix}0&0&0&0\\4&0&0&0\end{pmatrix},$$

$$\begin{pmatrix}2&1&0&0\\0&0&0&0\end{pmatrix}, \begin{pmatrix}1&1&0&0\\1&0&0&0\end{pmatrix}, \begin{pmatrix}0&1&0&0\\2&0&0&0\end{pmatrix}, \begin{pmatrix}2&0&0&0\\0&1&0&0\end{pmatrix}, \begin{pmatrix}1&0&0&0\\1&1&0&0\end{pmatrix},$$

$$\begin{pmatrix}0&0&0&0\\2&1&0&0\end{pmatrix}, \begin{pmatrix}1&0&1&0\\0&0&0&0\end{pmatrix}, \begin{pmatrix}0&0&1&0\\1&0&0&0\end{pmatrix}, \begin{pmatrix}1&0&0&0\\0&0&1&0\end{pmatrix}, \begin{pmatrix}0&0&0&0\\1&0&1&0\end{pmatrix},$$

$$\begin{pmatrix}0&2&0&0\\0&0&0&0\end{pmatrix}, \begin{pmatrix}0&1&0&0\\0&1&0&0\end{pmatrix}, \begin{pmatrix}0&0&0&0\\0&2&0&0\end{pmatrix}, \begin{pmatrix}0&0&0&1\\0&0&0&0\end{pmatrix}, \begin{pmatrix}0&0&0&0\\0&0&0&1\end{pmatrix}.$$

The types of $S_2 \sim S_4$ which characterize conjugacy classes which split over $S_2 \sim A_4$ are (cf. 1.27 (i)):

$$\begin{pmatrix}1&0&1&0\\0&0&0&0\end{pmatrix}, \begin{pmatrix}0&0&1&0\\1&0&0&0\end{pmatrix}, \begin{pmatrix}1&0&0&0\\0&0&1&0\end{pmatrix}, \begin{pmatrix}0&0&0&0\\1&0&1&0\end{pmatrix}.$$

The types of $S_2 \sim S_4$ which characterize conjugacy classes which split over $S_2 \sim {S_4}_{A_2}$ are (cf. 1.27 (ii)):

$$\begin{pmatrix}0&2&0&0\\0&0&0&0\end{pmatrix}, \begin{pmatrix}0&0&0&1\\0&0&0&0\end{pmatrix}.$$

The types of $S_2 \sim S_4$ which characterize conjugacy classes which split over $S_2 \sim {S_4}_{A_2}^{A_4}$ are (cf. 1.27 (iii)):

$$\begin{pmatrix}0&0&0&0\\0&2&0&0\end{pmatrix}, \begin{pmatrix}0&0&0&0\\0&0&0&1\end{pmatrix}.$$

2. Representations of wreath products and their characters

Let us now briefly recall what has been said in part I about the construction of matrix representations of wreath products $G \sim H$.

The salient point in connection with this construction was to notice that linear representations of the basis group can be extended, and how this can be done.

Let us assume that F is a linear representation of the group G over a field K with representation module M, underlying vector space V and a corresponding matrix representation $\mathbb{F}$.

For $n \in \mathbb{N}$, we form the n-fold tensor product of V with itself:

$$\overset{n}{\otimes} V := \underbrace{V \otimes_K \dots \otimes_K V}_{n\text{-times}} .$$

This vector space over K yields a left G^*-module, which we denote by $\overset{n}{\#} M$, if we define the operation of $(f;1_H) \in G^*$ on $\underset{i}{\otimes} v_i := v_1 \otimes \dots \otimes v_n \in \overset{n}{\otimes} V$ by

2.1 $$(f;1_H) \underset{i}{\otimes} v_i := \underset{i}{\otimes} f(i)v_i = f(1)v_1 \otimes \dots \otimes f(n)v_n .$$

The corresponding representation of G^* is denoted by $\overset{n}{\#} F$, a corresponding matrix representation $\overset{n}{\#} \mathbb{F}$ is defined by

2.2 $$\overset{n}{\#}\mathbb{F}(f;1_H) := \mathbb{F}(f(1)) \times \ldots \times \mathbb{F}(f(n))$$
$$= (f_{i_1k_1}(f(1)) \cdot \ldots \cdot f_{i_nk_n}(f(n))),$$

if $\mathbb{F}(g) = (f_{ik}(g))$, for $g \in G$.

The most important fact is now that this representation $\overset{n}{\#}F$ resp. $\overset{n}{\#}\mathbb{F}$ can be extended to a linear representation $\widetilde{\overset{n}{\#}}F$ resp. $\widetilde{\overset{n}{\#}}\mathbb{F}$ of $G \wr H$ (H a permutation group on $\mathbb{N}_n = \{1,\ldots,n\}$) in a natural way.

To do this we extend the set of operators on $\overset{n}{\otimes} V$ to the whole group $G \wr H$ as follows:

2.3 $$(f;\pi) \underset{i}{\otimes} v_i := \underset{i}{\otimes} f(i)v_{\pi^{-1}(i)} = f(1)v_{\pi^{-1}(1)} \otimes \ldots \otimes f(n)v_{\pi^{-1}(n)}.$$

This yields a left $G \wr H$-module with underlying vector space $\overset{n}{\otimes} V$, this module is denoted by $\widetilde{\overset{n}{\#}}M$. A corresponding matrix representation $\widetilde{\overset{n}{\#}}\mathbb{F}$ is obtained by applying suitable column permutations to the matrices $\overset{n}{\#}\mathbb{F}(f;1)$:

2.4 $$\widetilde{\overset{n}{\#}}\mathbb{F}(f;\pi) := (f_{i_1k_{\pi^{-1}(1)}}(f(1)) \ldots f_{i_nk_{\pi^{-1}(n)}}(f(n))).$$

Let us now evaluate the traces of these matrices. If m is the dimension of the vector space V and $\{b_1,\ldots,b_m\}$ a basis of V, then

$$\{\underset{\nu}{\otimes} b_{i_\nu} = b_{i_1} \otimes \ldots \otimes b_{i_n} \mid 1 \leq i_\nu \leq m\}$$

is a basis of $\overset{n}{\otimes} V$.

Then 2.3 yields

$$(f;(1\ldots n)) \underset{\nu}{\otimes} b_{i_\nu} = \underset{\nu}{\otimes} f(\nu) b_{i_{(1\ldots n)^{-1}(\nu)}}$$

$$= (\sum_i f_{ii_n}(f(1))b_i) \otimes (\sum_j f_{ji_1}(f(2))b_j) \otimes \ldots \otimes (\sum_k f_{ki_{n-1}}(f(n))b_k)$$

$$= \sum_{1\leq i,j,\ldots,k\leq m} f_{ii_n}(f(1)) f_{ji_1}(f(2)) \ldots f_{ki_{n-1}}(f(n)) b_i \otimes \ldots \otimes b_k .$$

This yields for the trace:

2.5 $\operatorname{tr} \widetilde{\#^n \mathbb{F}}(f;(1\ldots n)) = \sum_{1\leq i_\nu \leq m} f_{i_1 i_n}(f(1)) f_{i_n i_{n-1}}(f(n)) \ldots f_{i_2 i_1}(f(2))$

$$= \operatorname{tr} \mathbb{F}(f(1)f(n)\ldots f(2))$$

$$= \operatorname{tr} \mathbb{F}(g),$$

where g denotes the cycle product associated with $(1\ldots n)$ with respect to f.

In order to obtain the trace of $\widetilde{\#^n \mathbb{F}}(f;\pi)$ for a general element $\pi \in S_n$, we need only notice that 2.3 says that $(e;\pi)$ acts on $\underset{i}{\otimes} v_i$ by just permuting cyclically the factors correspon-ing to the symbols in the cycles. This yields for the character $\chi^{\widetilde{\#^n F}}$ of $\widetilde{\#^n F}$:

2.6 If $(f;\pi) \in G\wr H$, $T(f;\pi) = (a_{ik}(f;\pi))$, $\pi = \prod_{\nu=1}^{c(\pi)} \pi_\nu$

$$= \prod_{\substack{i,j,k \\ a_{ik}(f;\pi)>0}} \pi_{ik}^j \quad \text{(cf. 1.9), and} \quad b_i(f;\pi) := \sum_k a_{ik}(f;\pi),$$

then

$$\chi^{\widetilde{n}\# F}(f;\pi) = \prod_{\nu=1}^{c(\pi)} \chi^F(g_\nu(f;\pi)) = \prod_{\substack{i,j,k\\ a_{ik}(f;\pi)>0}} \chi^F(g_{ik}^j(f;\pi))$$

$$= \prod_{\substack{i\\ a_{ik}(f;\pi)>0}} (\chi_i^F)^{b_i(f;\pi)} ,$$

where χ_i^F denotes the value of χ^F on the conjugacy class C^i of G.

A corollary of 2.6 is:

2.7 (i) $\forall\, g \in G,\ \pi \in H\ (\chi^{\widetilde{n}\# F}(g,\ldots,g;\pi) = \prod_{k=1}^{n} \chi^F(g^k)^{a_k(\pi)})$,

(ii) $\forall\, \pi \in H\ (\chi^{\widetilde{n}\# F}(e;\pi) = \chi^F(1)^{c(\pi)} = (f^F)^{c(\pi)})$,

(iii) $\forall\, g \in G\ (\chi^{\widetilde{n}\# F}(g,\ldots,g;1) = \chi^F(g)^n)$.

Later on, 2.7 will turn out to be very useful.

We proceed quite analogously when we start from a representation F^* of the basis group G^* which is a product of several maybe different representations $F', F'', \ldots$ of G, say:

$$F^* := \underbrace{F' \# \ldots \# F'}_{n'\text{-times}} \# \underbrace{F'' \ldots \# F''}_{n''\text{-times}} \# \ldots$$

If $S_{n'}$, $S_{n''}$, ... denote the subgroups of S_n which consist of the $n'!$, $n''!$, ... permutations which move at most the first n' symbols, the next n'' symbols, ..., then F^* can be extended to $G \backslash (H \cap S_{n'} \times S_{n''} \times \ldots)$ quite analogously.

I applied this in part I to the case, where the factors of F^* are irreducible representations over an algebraically closed field K, assuming G to be a finite group. Let me recall this.

Let $F^1, \ldots, F^r$ denote a complete system of pairwise inequivalent and irreducible representations of G over K with corresponding representation modules $M^1, \ldots, M^r$, underlying vector spaces $V^1, \ldots, V^r$ and corresponding matrix representations $\mathbb{F}^1, \ldots, \mathbb{F}^r$.

Then each irreducible K-representation of G^* is of the form

$$F^* := F_1 \# \ldots \# F_n =: \mathop{\#}_i F_i, \quad \text{where} \quad F_i \in \{F^1, \ldots, F^r\},$$

with representation module

$$M^* := M_1 \# \ldots \# M_n =: \mathop{\#}_i M_i, \quad \text{where} \quad M_i := M^j, \quad \text{if} \quad F_i = F^j.$$

The underlying vector space is

$$V^* := V_1 \otimes_K \ldots \otimes_K V_n =: \mathop{\otimes}_i V_i \ ,$$

where V_i denotes the underlying vector space of M_i.

If n_j denotes the number of factors F_i of F^*, which are equal to F^j, for $1 \leq j \leq r$, then

$$TF^* := (n_1,\ldots,n_r) =: (n)$$

was called <u>the type of</u> F^*.

Denoting then by S_{n_j} the subgroup of $S_n(\geq H)$ consisting of the permutations, which move at most the indices i of factors F_i equal to F^j, if $n_j > 0$, or $S_{n_j} := \{1_{S_n}\}$, if $n_j = 0$, we obtained for the inertia group $G\sim H_{F^*}$ of F^* in $G\sim H$ (cf. I. 5.10):

2.8 $$\underline{G\sim H_{F^*} := \{(f;\pi) \mid F^{*(f;\pi)} \text{ equivalent } F^*\}}$$

$$\underline{= \{(f;\pi) \mid \forall\, i\ (F_{\pi(i)} = F_i)\}}$$

$$\underline{= G^*(H \cap S_{n_1}\times\ldots\times S_{n_r})' = G\sim(H \cap S_{(n)})},$$

where $S_{(n)} := S_{n_1} \times\ldots\times S_{n_r}$.

F^* can be extended to a linear representation $\widetilde{F^*}$ of $G\sim H_{F^*}$ as has been described above in 2.3:

2.9 $$\forall\ (f;\pi) \in G\sim H_{F^*}\ \left((f;\pi) \underset{i}{\otimes} v_i := \underset{i}{\otimes} f(i)v_{\pi^{-1}(i)}\right),$$

for $(f;\pi) \in G\sim H_{F^*}$ implies $F_{\pi(i)} = F_i$, so that $f(i)v_{\pi^{-1}(i)}$ is defined.

In this way we obtain a left $G\sim H_{F^*}$-module $\widetilde{M^*}$ with underlying vector space V^*. If

$$\mathbf{F}_j(g) = (F^j_{ik}(g)),\ 1 \leq j \leq n,$$

then a corresponding matrix representation is

$$\widetilde{\mathbf{F}^*}(f;\pi) = (F^1_{i_1 k_{\pi^{-1}(1)}}(f(1)) \dots F^n_{i_n k_{\pi^{-1}(n)}}(f(n))). \qquad \underline{\underline{2.10}}$$

$\widetilde{F^*}$ is irreducible since its restriction to G^* is just F^*.

$$\widetilde{F^*} \downarrow G^* = F^*. \qquad \underline{\underline{2.11}}$$

For the character of $\widetilde{F^*}$ we obtain from 2.6:

<u>2.12</u> <u>If $(f;\pi) \in G \wr H_{F^*}$, where $\pi = \prod_{\substack{i,j,k \\ a_{ik}(f;\pi)>0}} \pi^j_{ik}$</u>

<u>and $g^j_{ik}(f;\pi)$ is the cycle product associated with</u>

<u>$\pi^j_{ik} = (r^j_{ik} \dots \pi^{k-1}(r^j_{ik}))$, then</u>

$$\chi^{\widetilde{F^*}}(f;\pi) = \prod_{\substack{i,j,k \\ a_{ik}(f;\pi)>0}} \chi^{F_{r^j_{ik}}}(g^j_{ik}(f;\pi)).$$

This formula was stated by Klaiber (Klaiber [1]) without proof.

If in addition to such an irreducible representation $\widetilde{F^*}$ of $G \wr H_{F^*} = G \wr (H \cap S_{(n)})$, F'' is an irreducible K-representation of $H \cap S_{(n)}$, we obtain a second irreducible K-representation F' of $G \wr H_{F^*}$ by putting

2.13 $$F'(f;\pi) := F''(\pi).$$

Then the product

$$\widetilde{F^*} \otimes F'$$

is a third irreducible K-representation of $G \setminus H_{F^*}$.

Clifford's theory of representations of groups with normal subgroups yields that each irreducible K-representation of $G \setminus H$ is of the form

2.14 $$F := (\widetilde{F^*} \otimes F') \uparrow G \setminus H.$$

This theory also yields a result how we can obtain just a complete system of irreducible K-representations of $G \setminus H$ (cf. I 5.20):

2.15 If G is a finite group and K an algebraically closed field, then $F = (\widetilde{F^*} \otimes F') \uparrow G \setminus H$ runs exactly through a complete system of pairwise inequivalent and irreducible K-representations F of $G \setminus H$, if F^* runs through a complete system of pairwise not conjugate (with respect to H') but irreducible K-representations of G^*, and, while F^* is fixed, F'' (cf. 2.13) runs through a complete system of

<u>pairwise inequivalent and irreducible K-represen-tations of the inertia factor $H \cap S_{(n)}$ of F^*.</u>

In order to construct the irreducible K-representations of $G \sim H$ we need only know the representations of G and of the subgroups $H \cap S_{(n)} = H \cap S_{n_1} \times \ldots \times S_{n_r}$ of H.
All properties which remain valid under inner tensor product multiplication and induction, hold for the representations of $G \sim H$, if they hold for the representations of G and $H \cap S_{(n)}$ (cf. I 5.39 - 5.42). This together with 2.12 also yields results about characters, e.g.

2.16 <u>If the ordinary irreducible characters of G as well as of the subgroups $H \cap S_{(n)}$ of H are rational integral (real), then all the ordinary irreducible characters of $G \sim H$ are rational integral (real).</u>

A corollary is I 3.14, which we now obtain without referring to conjugacy classes: If G is ambivalent, $G \sim S_n$ is ambivalent as well. For the ambivalency of a finite group is equivalent to the reality of its characters. We also obtain I 3.16 again: The ambivalency of $G \sim H$ implies the ambivalency of G and the ambivalency of H.

Since the ordinary irreducible characters of symmetric groups are rational integral-valued, we obtain as a special case of 2.16:

2.17 If the character table of G is rational integral, then this holds for the character table of $G\sim S_n$.

Besides these more theoretical results 2.16 and 2.17, 2.12 together with 2.15 allow the evaluation of the character table of $G\sim H$, when the character tables of G as well as of the subgroups $H \cap S_{(n)}$ of H are known.

2.12 yields for the character of $\widetilde{F^*} \otimes F'$:

2.18 $$\forall (f;\pi) \in G\sim H_{F^*} \left(\chi^{\widetilde{F^*}\otimes F'}(f;\pi) = \chi^{F''}(\pi) \prod_{\substack{i,j,k \\ a_{ik}(f;\pi)>0}} \chi^{F_{r_{ik}^j}}(g_{ik}^j(f;\pi))\right),$$

so that furthermore only an implementation of the inducing process is needed. Gretschel and Hilge (Gretschel/Hilge [1]) have done this, extending the results of Sänger (Sänger [1]) so that now the character tables of $S_m \sim S_n$ are known for $mn \leq 15$.

If F_G denotes a representation of G and F_H a representation of H, I introduced in part I the following abbreviation

$$(F_G;F_H) := \widetilde{\#^n F_G} \otimes F_H.$$

2.18 yields for its character:

2.19 $\forall (f;\pi) \in G\setminus H \ (\chi^{(F_G;F_H)} = \chi^{F_H}(\pi) \prod_i (\chi_i^{F_G})^{b_i(f;\pi)})$.

This formula has been used by Littlewood (Littlewood [3], [4]), I shall revert to this later on.

Let us continue this section with a numerical example.

Let us consider the splitting of ordinary irreducible representations of hyperoctahedral groups $S_2\setminus S_n$ over $S_2\setminus A_n$ and $S_2\setminus S_{n_{A_2}}$ which corresponds to the splitting of conjugacy classes described in the preceding section (for more details and for the splitting over $S_2\setminus S_{n_{A_2}}^{A_n}$ cf. Celik / Kerber / Pahlings [1]). Since $\{[2], [1^2]\}$ is a complete system of pairwise inequivalent and irreducible representations of S_2 over $\mathbb{C}$ (cf. I 4.27), the following is a complete system of ordinary irreducible representations of pairwise different types of the basis group $S_2{}^*$ of $S_2\setminus S_n$:

2.20 $\{\overset{s}{\#}[2]\overset{t}{\#}[1^2] := \underbrace{[2]\#\ldots\#[2]}_{s\text{-times}}\#\underbrace{[1^2]\#\ldots\#[1^2]}_{t\text{-times}} \mid s,t\in\mathbb{Z}_{\geq 0} \wedge s+t=n\}$.

The inertia group of $\overset{s}{\#}[2]\overset{t}{\#}[1^2]$ is $S_2{}^*(S_s\times S_t)' = S_2\setminus(S_s\times S_t)$, hence each ordinary irreducible representation F'' of the inertia factor $S_s \times S_t$ is of the form $[\alpha]\#[\beta]$, where α denotes a partition of s (for short: $\alpha \vdash s$), and β a partition of t ($\beta \vdash t$) (if $s = 0$, we put $[\alpha]\#[\beta] = [0]\#[\beta] := [\beta]$, and correspondingly $[\alpha]\#[\beta] := [\alpha]$, if $t = 0$).

The ordinary irreducible representation of $S_2 \sim S_n$ which arises is denoted by

2.21 $$(\alpha|\beta) := (\overbrace{\# [2] \# [1^2]}^{s \quad t} \otimes ([\alpha]\#[\beta])')\uparrow S_2 \sim S_n.$$

This together with 2.15 yields Young's result:

2.22 A complete system of pairwise inequivalent ordinary irreducible representations of $S_2 \sim S_n$ is

$$\{(\alpha|\beta) \mid s,t \in \mathbb{Z}_{\geq 0} \wedge s+t=n \wedge \alpha \vdash s \wedge \beta \vdash t\}.$$

For an example I mention that the following is a complete system of pairwise inequivalent ordinary irreducible representations of $S_2 \sim S_4$:

2.23 $$\{(4|0),(0|4),(3,1|0),(0|3,1),(2^2|0),(0|2^2),(2,1^2|0),(0|2,1^2),$$
$$(1^4|0),(0|1^4),(3|1),(1|3),(2,1|1),(1|2,1),(1^3|1),(1|1^3),$$
$$(2|2),(2|1^2),(1^2|2),(1^2|1^2)\}.$$

This system corresponds to the system of conjugacy classes of $S_2 \sim S_4$ given in the preceding section.

We ask now for a complete system of ordinary irreducible

representations of $S_2 \wr A_n$ and $S_2 \wr S_{n_{A_2}}$.

Clifford's theory of representations of groups with normal subgroups yields that for this we need only show, which of the representations $(\alpha|\beta)$ of $S_2 \wr S_n$ form pairs of associated representations with respect to the normal subgroups of index 2 considered (so that their restrictions to the normal subgroup are equal and irreducible). The remaining selfassociated representations split into two conjugate and irreducible representations when restricted to the normal subgroup, these two irreducible constituents have to be described precisely, say as representations induced by certain representations of subgroups of the normal subgroup.

If we are given $(\alpha|\beta)$ of $S_2 \wr S_n$, then we obtain the associated representation by forming its tensor product with the alternating representation of $S_2 \wr S_n$ with respect to the normal subgroup considered.

The following is obviously valid:

2.24 <u>The alternating representations of $S_2 \wr S_n$ with respect to $S_2 \wr A_n$, $S_2 \wr S_{n_{A_2}}$, $S_2 \wr S_{n_{A_2}}^{A_n}$, respectively, are the representations $(1^n|0)$, $(0|n)$, $(0|1^n)$, respectively.</u>

Forming the inner tensor products we obtain the desired associated and selfassociated representations:

2.25 (i) The ordinary irreducible representations of $S_2 \sim S_n$ which are associated with $(\alpha|\beta)$ with respect to $S_2 \sim A_n$, $S_2 \sim S_{n_{A_2}}$, $S_2 \sim S_{n_{A_2}}^{A_n}$, respectively, are:

$$(\alpha'|\beta') = (\alpha|\beta) \otimes (1^n|0),$$

$$(\beta|\alpha) = (\alpha|\beta) \otimes (0|n),$$

$$(\beta'|\alpha') = (\alpha|\beta) \otimes (0|1^n).$$

(ii) The ordinary irreducible representation $(\alpha|\beta)$ of $S_2 \sim S_n$ is selfassociated with respect to $S_2 \sim A_n$, $S_2 \sim S_{n_{A_2}}$, $S_2 \sim S_{n_{A_2}}^{A_n}$, respectively, if and only if

$$\alpha = \alpha' \wedge \beta = \beta',$$

$$\alpha = \beta,$$

$$\alpha = \beta',$$

respectively.

(Recall from part I, p. 20, that α' denotes the partition which is associated with α.)

For example

2.26 (i) The pairs of ordinary irreducible representations of $S_2 \sim S_4$ which are associated with respect to $S_2 \sim A_4$ are:

$\{(4|0),(1^4|0)\},\{(0|4),(0|1^4)\},\{(2,1^2|0),(3,1|0)\},$
$\{(0|2,1^2),(0|3,1)\},\{(3|1),(1^3|1)\},\{(1|3),(1|1^3)\},$
$\{(2|2),(1^2|1^2)\},\{(2|1^2),(1^2|2)\}.$

The ordinary irreducible representations of $S_2 \sim S_4$ which are selfassociated with respect to $S_2 \sim A_4$ are:

$$(2^2|0),(0|2^2),(2,1|1),(1|2,1).$$

(ii) The pairs of ordinary irreducible representations of $S_2 \sim S_4$ which are associated with respect to $S_2 \sim S_{4_{A_2}}$ are:

$\{(4|0),(0|4)\},\{(3,1|0),(0|3,1)\},\{(2^2|0),(0|2^2)\},$
$\{(2,1^2|0),(0|2,1^2)\},\{(1^4|0),(0|1^4)\},\{(3|1),(1|3)\},$
$\{(2,1|1),(1|2,1)\},\{(1^3|1),(1|1^3)\},\{(2|1^2),(1^2|2)\}.$

The ordinary irreducible representations of $S_2 \sim S_4$ which are selfassociated with respect to $S_2 \sim S_{4_{A_2}}$ are:

$$(2|2),(1^2|1^2).$$

(iii) The pairs of ordinary irreducible representations of $S_2 \sim S_4$ which are associated with respect to $S_2 \sim S_{4_{A_2}}^{A_4}$ are:

$\{(4|0),(0|1^4)\},\{(0|4),(1^4|0)\},\{(3,1|0),(0|2,1^2)\},$

$\{(0|3,1),(2,1^2|0)\},\{(2^2|0),(0|2^2)\},\{(3|1),(1|1^3)\},$

$\{(1|3),(1^3|1)\},\{(2,1|1),(1|2,1)\},\{(2|2),(1^2|1^2)\}.$

The ordinary irreducible representations of $S_2 \sim S_4$ which are selfassociated with respect to $S_2 \sim S_{4_{A_2}}^{A_4}$ are:

$$(2|1^2),(1^2|2).$$

As has been mentioned above, the elements of a pair of associated representations restrict to the same irreducible representation of the normal subgroup of index 2, while restrictions of selfassociated representations of $S_2 \sim S_n$ split into two representations.

This yields various symmetry properties of the character table of $S_2 \sim S_n$, some of which have been described in part I (cf. I 6.13 - 6.15). It allows also to derive a great deal of the character table of $S_2 \sim A_n$, $S_2 \sim S_{n_{A_2}}$ and $S_2 \sim S_{n_{A_2}}^{A_n}$ from the character table of $S_2 \sim S_n$.

But we are also interested in the remaining problem: What are the remaining entries of the character table of the normal subgroup considered? i.e. we would like to know the values of characters of the irreducible constituents of the restrictions of the selfassociated representations of $S_2 \sim S_n$.

An answer to this question would imply an answer to the question: are the values of the characters of $S_2 \wr A_n$, $S_2 \wr S_{n_{A_2}}$, $S_2 \wr S_{n_{A_2}}^{A_n}$ rational integral?

The values of the characters of $S_2 \wr A_n$ are not rational integral in general, this is obvious. For example, $S_2 \wr A_3$ has even a strictly complex character table: The irreducible constituents $[2,1]^{\pm}$ of $[2,1] \downarrow A_3$ (cf. I 4.54) yield the irreducible representations

$$\widetilde{\#^{3}[2]} \otimes [2,1]^{\pm'} = (2;2,1^{\pm})$$

of $S_2 \wr A_3$ with strictly complex-valued characters.
Nevertheless, the characters of $S_2 \wr S_{n_{A_2}}$ and $S_2 \wr S_{n_{A_2}}^{A_n}$ may have rational integral values. It is in fact known that Weyl groups of type D_n have character tables over $\mathbb{Z}$, even that $\mathbb{Q}$ is a splitting field for these (cf. Curtis/Benson [1]).
I would like to show this directly by constructing a complete system of pairwise inequivalent and irreducible ordinary representations of these groups.
For this we need only construct the constituents of the restrictions of selfassociated representations of $S_2 \wr S_n$. We start with the normal subgroup $S_2 \wr A_n$, so that we have to show how the two irreducible constituents of the representation

2.27 $(\alpha|\beta)\downarrow S_2\wr A_n$, where $\alpha = \alpha'$ and $\beta = \beta'$,

can be obtained as representations induced from certain representations of a suitable subgroup of $S_2\wr A_n$.

To do this we first apply 2.15 to the wreath product $S_2\wr A_n$ in order to construct a complete system of its ordinary irreducible representations. We then ask which of these irreducible representations are the irreducible constituents of 2.27.

The ordinary irreducible representations of the basis group $S_2{}^*$ of $S_2\wr A_n$ are again the representations

2.28 $F^* := \overset{s}{\#}[2]\overset{t}{\#}[1^2]$, where $s,t \in \mathbb{Z}_{\geq 0}$ and $s+t=n$.

The inertia group of F^* is

2.29 $S_2\wr(S_s\times S_t\cap A_n)$,

so that we need to know the ordinary irreducible representations of $S_s \times S_t \cap A_n$.

From the complete system

2.30 $\{[\alpha]\#[\beta] \mid \alpha\vdash s \wedge \beta\vdash t\}$

of ordinary irreducible representations of $S_s \times S_t$, we obtain at once a complete system of representations of $S_s \times S_t \cap A_n$ if s or t is less than or equal to 1, for then $S_s \times S_t \cap A_n$ is A_s or A_t, so that we need only apply I 4.54. Let us exclude these trivial cases by assuming that $s,t \geq 2$.

We may then consider the series

2.31 $$A_s \times A_t \underset{2}{\triangleleft} S_2 \times S_t \cap A_n \underset{2}{\triangleleft} S_s \times S_t.$$

It is obvious, that exactly the elements

2.32 $$[\alpha]\#[\beta], \text{ where } \alpha = \alpha' \text{ and } \beta = \beta',$$

of 2.30 are selfassociated with respect to $S_s \times S_t \cap A_n$.

In order to obtain the irreducible constituents of the restriction

2.33 $$[\alpha=\alpha']\#[\beta=\beta'] \downarrow S_s \times S_t \cap A_n$$

of the representation 2.32 we use 2.31 and notice that the following is valid (recall that $s,t \geq 2$):

2.34 $$\underline{[\alpha=\alpha']\#[\beta=\beta'] \downarrow A_s \times A_t = [\alpha]^+\#[\beta]^+ + [\alpha]^+\#[\beta]^- + [\alpha]^-\#[\beta]^+ + [\alpha]^-\#[\beta]^-.}$$

Since everyone of these four irreducible constituents induces $[\alpha]\#[\beta]$ in $S_s \times S_t$, Frobenius' reciprocity law implies that two of them induce the irreducible constituents of 2.33 in $S_s \times S_t \cap A_n$.

A consideration of the representing matrices then shows the following equivalences (use I 4.55):

2.35 $$\underline{[\alpha]^+\#[\beta]^+ \uparrow S_s \times S_t \cap A_n \sim [\alpha]^-\#[\beta]^- \uparrow S_s \times S_t \cap A_n,}$$

$$\underline{[\alpha]^+\#[\beta]^- \uparrow S_s \times S_t \cap A_n \sim [\alpha]^-\#[\beta]^+ \uparrow S_s \times S_t \cap A_n.}$$

Hence we can conclude that the following holds:

2.36 If $n=s+t$, where $s,t,\geq 2$, and $\alpha=\alpha' \vdash s, \beta=\beta' \vdash t$,

then

$$[\alpha]\#[\beta] \downarrow S_s\times S_t\cap A_n = [\alpha]^+\#[\beta]^+\uparrow S_s\times S_t\cap A_n + [\alpha]^+\#[\beta]^-\uparrow S_s\times S_t\cap A_n$$

is the decomposition of the restriction $[\alpha]\#[\beta]\downarrow S_s\times S_t\cap A_n$ into its irreducible constituents.

This allows us to construct a complete system of ordinary irreducible representations of the inertia factor of the representation 2.28 in $S_2\sim A_n$.

In order to obtain just a complete system of ordinary irreducible representations of $S_2\sim A_n$, we would like to know which representations F^* has to run through. Since F^* consists of factors of just two kinds, namely $[2]$ and $[1^2]$, it suffices to notice that A_n is $(n-2)$-fold transitive. This yields that for $n \geq 4$ (so that $n-(n-2) \leq n-2$) F^* needs only to run through a complete system of irreducible representations of $S_2{}^*$ of different types. The cases $n = 1,2,3$ are easy to handle.

Hence we are left with the question, which of these irreducible representations of $S_2\sim A_n$ are the irreducible constituents of

2.37 $$(\alpha=\alpha'|\beta=\beta') \downarrow S_2\sim A_n.$$

Since the matrices of $(\alpha|\beta)$ are up to a sign for each nonzero box just the matrices of

$$[\alpha][\beta] = [\alpha]\#[\beta] \uparrow S_n = [\alpha]^+\#[\beta]^+ \uparrow S_n = [\alpha]^+\#[\beta]^- \uparrow S_n$$

(cf. I 5.24), we see that since 2.36 holds, the desired irreducible constituents of 2.37 are just

$$(\overbrace{\overset{s}{\#}[2]\overset{t}{\#}[1^2]}\otimes([\alpha]^+\#[\beta]^+ \uparrow S_s\times S_t\cap A_n)') \uparrow S_2\wr A_n =: (\alpha^+|\beta^+) \uparrow S_2\wr A_n$$

and

$$(\overbrace{\overset{s}{\#}[2]\overset{t}{\#}[1^2]}\otimes([\alpha]^+\#[\beta]^- \uparrow S_s\times S_t\cap A_n)') \uparrow S_2\wr A_n =: (\alpha^+|\beta^-) \uparrow S_2\wr A_n$$

if s and $t > 1$. If $t=0$ or $t=1$, they are $(\alpha^+|0) \uparrow S_2\wr A_n$ and $(\alpha^-|0) \uparrow S_2\wr A_n$, $(\alpha^+|1) \uparrow S_2\wr A_n$ and $(\alpha^-|1) \uparrow S_2\wr A_n$ and analogously if $s=0$ or $s=1$: $(0|\beta^+) \uparrow S_2\wr A_n$ and $(0|\beta^-) \uparrow S_2\wr A_n$, $(1|\beta^+) \uparrow S_2\wr A_n$ and $(1|\beta^-) \uparrow S_2\wr A_n$.

For example:

$$(2^2|0) \downarrow S_2\wr A_4 = (2^{2+}|0) + (2^{2-}|0),$$

$$(0|2^2) \downarrow S_2\wr A_4 = (0|2^{2+}) + (0|2^{2-}),$$

$$(2,1|1) \downarrow S_2\wr A_4 = (2,1^+|1) + (2,1^-|1),$$

$$(1|2,1) \downarrow S_2\wr A_4 = (1|2,1^+) + (1|2,1^-).$$

Thus, the following system is a complete system of pairwise inequivalent ordinary irreducible representations of $S_2\wr A_4$:

2.38 $\{(4|0)\downarrow S_2\sim A_4,\ (0|4)\downarrow S_2\sim A_4,\ (2,1^2|0)\downarrow S_2\sim A_4,$
$(0|2,1^2)\downarrow S_2\sim A_4,\ (3|1)\downarrow S_2\sim A_4,\ (1|3)\downarrow S_2\sim A_4,$
$(2|2)\downarrow S_2\sim A_4,\ (2|1^2)\downarrow S_2\sim A_4,\ (2^{2+}|0),\ (2^{2-}|0),$
$(0|2^{2+}),\ (0|2^{2-}),\ (2,1^+|1),\ (2,1^-|1),\ (1|2,1^+),\ (1|2,1^-)\}.$

These results together with the results of section 1 about the conjugacy classes of $S_2\sim A_n$ allow the evaluation of the character table of $S_2\sim A_n$ (which in general contains complex numbers).

What can be said about the characters of $S_2\sim S_{n_{A_2}}$?

It is known, that their values are rational integral. I would like to derive this, using the above results. 2.25 says that we need only consider the representations $(\alpha|\alpha)$ of $S_2\sim S_n$ and to obtain the irreducible constituents of its restriction

2.39 $$(\alpha|\alpha)\downarrow S_2\sim S_{n_{A_2}}$$

say as representations induced by certain representations of suitable subgroups.

To do this, we apply Clifford's theory to the normal subgroup

2.40 $$S_2{*}\cap S_2\sim S_{n_{A_2}} = \{(f;1_{S_n}) \mid \prod_i f(i) = 1_{S_2}\} \leq S_2\sim S_{n_{A_2}}.$$

Every ordinary irreducible representation of this group is of the form

2.41 $$\overset{s}{\#}[2]\overset{t}{\#}[1^2]\downarrow S_2{*}\cap S_2\sim S_{n_{A_2}}.$$

Since $(\alpha|\alpha)$ arises from

$$\overset{s}{\#}[2]\overset{s}{\#}[1^2],\ s := \frac{n}{2}\ ,$$

we need only consider, which of the irreducible representations of $S_2 \sim S_{n_{A_2}}$ arise from

2.42 $$\overset{s}{\#}[2]\overset{s}{\#}[1^2] \downarrow S_2{*} \cap S_2 \sim S_{n_{A_2}} .$$

The inertia group of this representation 2.42 consists of all the $(f;\pi) \in S_2 \sim S_{n_{A_2}}$ which satisfy for each $(f';1) \in S_2{*} \cap S_2 \sim S_{n_{A_2}}$:

2.43 $$\operatorname{sgn} \prod_{s+1}^{n} f'(j) = \overset{s}{\#}[2]\overset{s}{\#}[1^2](f';1) = \overset{s}{\#}[2]\overset{s}{\#}[1^2](f;\pi)(f';1)(f;\pi)^{-1}$$
$$= \operatorname{sgn} \prod_{s+1}^{n} f'_{\pi}(j).$$

This holds if and only if $\pi \in \varphi[S_s \sim S_2] \leq S_n$. Hence we obtain

<u>2.44</u> <u>If n is even and $s := \frac{n}{2}$, the inertia group of $\overset{s}{\#}[2]\overset{s}{\#}[1^2] \downarrow S_2{*} \cap S_2 \sim S_{n_{A_2}}$ in $S_2 \sim S_{n_{A_2}}$ is</u>

<u>$$(S_2{*} \cap S_2 \sim S_{n_{A_2}})\varphi[S_s \sim S_2]' = S_2 \sim (\varphi[S_s \sim S_2])_{A_2} .$$</u>

The restriction of $\overset{s}{\#}[2]\overset{s}{\#}[1^2]$ to $S_2^* \cap S_2 \sim S_{n_{A_2}}$ can be extended to $S_2 \sim (\varphi[S_s \sim S_2])_{A_2}$ as follows:

$$\widetilde{\overset{s}{\#}[2]\overset{s}{\#}[1^2]}\,(f;(g;\rho)) := \prod_{i=s+1}^{n} \operatorname{sgn}(f(i)).$$

The check is very easy.

The inertia factor is

$$(S_2 \sim (\varphi[S_s \sim S_2])_{A_2} \Big/ (S^* \cap S_2 \sim S_{n_{A_2}}) \simeq S_s \sim S_2,$$

so that the above extension produces the two irreducible representations

$$\widetilde{\overset{s}{\#}[2]\overset{s}{\#}[1^2]} \otimes (\alpha;2)'$$

and

$$\widetilde{\overset{s}{\#}[2]\overset{s}{\#}[1^2]} \otimes (\alpha;1^2)'$$

of the inertia group $S_2 \sim (\varphi[S_s \sim S_2])_{A_2}$.

We notice that

$$(\alpha;2) = \widetilde{[\alpha]\#[\alpha]} \otimes [2]', \quad (\alpha;1^2) = \widetilde{[\alpha]\#[\alpha]} \otimes [1^2]'$$

yields

$$(\widetilde{\overset{s}{\#}[2]\overset{s}{\#}[1^2]} \otimes ([\alpha]\#[\alpha])') \uparrow S_2 \sim (\varphi[S_s \sim S_2]) \downarrow S_2 \sim (\varphi[S_s \sim S_2])_{A_2}$$

$$= \underbrace{\widetilde{\overset{s}{\#}[2]\overset{s}{\#}[1^2]}} \otimes ((\alpha;2) + (\alpha;1^2))'$$

extended to $S_2 \sim (\varphi[S_s \sim S_2])_{A_2}$.

We have obtained

2.45 If n is even, $s := \frac{n}{2}$, and $\alpha \vdash \frac{n}{2}$, then the decomposition of $(\alpha|\alpha) \downarrow S_2 \sim S_{n_{A_2}}$ into its irreducible constituents is

$$(\alpha|\alpha) \downarrow S_2 \sim S_{n_{A_2}} = (\alpha|\alpha)_+ + (\alpha|\alpha)_- \text{ , where}$$

$$(\alpha|\alpha)_+ := (\widetilde{\overset{s}{\#}[2]\overset{s}{\#}[1^2] \downarrow S_2{}^* \cap S_2 \sim S_{n_{A_2}}} \otimes (\alpha;2)') \uparrow S_2 \sim S_{n_{A_2}} ,$$

$$(\alpha|\alpha)_- := (\widetilde{\overset{s}{\#}[2]\overset{s}{\#}[1^2] \downarrow S_2{}^* \cap S_2 \sim S_{n_{A_2}}} \otimes (\alpha;1^2)') \uparrow S_2 \sim S_{n_{A_2}} .$$

This together with 2.25 yields:

2.46 Every ordinary representation of $S_2 \sim S_{n_{A_2}}$ can be written over $\mathbb{Q}$ (and hence over $\mathbb{Z}$), so that especially the characters of these groups, i.e. of the Weyl groups of type D_n have rational integral values. Hence the character table of each member of the series of Weyl groups is rational integral.

This result was obtained by Benson and Curtis in a different way (Benson/Curtis [1]). The preceding direct derivation can be generalized (Celik/Pahlings/Kerber [1]) so that we obtain e.g. the ordinary irreducible representations of $G \sim S_{n_M}$ and $G \sim S_{n_M}^{A_n}$,

M a subgroup of index 2 in G, G abelian.

The preceding theorems on the splitting of conjugacy classes and of ordinary irreducible representations can be applied to modular representation theory, too. For if we have evaluated the decomposition matrix of $S_2 \wr S_n$, say, we may ask for the decomposition matrices of $S_2 \wr A_n$, $S_2 \wr S_{n_{A_2}}$ and $S_2 \wr S_{n\;A_2}^{A_n}$, for which we need to know the splitting of both ordinary irreducible representations and p-regular conjugacy classes. Let us consider an example: $S_2 \wr S_4$,p := 2.

In part I the decomposition matrices of S_2 and S_4 with respect to p = 2 were evaluated, we obtained (I, 7.12, 7.16):

$$\underline{\underline{2.47}} \qquad \begin{array}{c} \hat{F}_1 \\ \left[\begin{array}{c} 1 \\ 1 \end{array}\right] \end{array} \begin{array}{l} \\ [2] \\ [1^2] \end{array} \qquad\qquad \begin{array}{cc} \check{F}_1 & \check{F}_2 \\ \multicolumn{2}{c}{\left[\begin{array}{cc} 1 & 0 \\ 1 & 1 \\ 0 & 1 \\ 1 & 1 \\ 1 & 0 \end{array}\right]} \end{array} \begin{array}{l} \\ [4] \\ [3,1] \\ [2^2] \\ [2,1^2] \\ [1^4] \end{array}$$

Hence there are exactly two 2-modular irreducible representations of $S_2 \wr S_4$, namely

$$\underline{\underline{2.48}} \qquad F_1 := (\hat{F}_1;\check{F}_1) \quad \text{and} \quad F_2 := (\hat{F}_1;\, \check{F}_2)$$

In order to obtain the decomposition matrix of $S_2 \wr S_4$ we need evaluate the multiplicity of F_i in $(\alpha|\beta)$, a 2-modular representation of $S_2 \sim S_4$ corresponding to $(\alpha|\beta)$. Since (recall $p = 2$):

$$\overline{(\alpha|\beta)} = \overline{([\alpha]\#[\beta])'} \uparrow S_2 \sim S_4$$

we obtain

$$\forall (f;\pi) \in S_2 \sim S_4 \ (\overline{(\alpha|\beta)}(f;\pi) = \overline{[\alpha][\beta]}\ (\pi)),$$

so that the following holds for the multiplicity of F_i in $\overline{(\alpha|\beta)}$:

2.49 $\quad (\overline{(\alpha|\beta)}, F_i) = (\overline{[\alpha][\beta]}, \check{F}_i), \quad i = 1,2.$

The multiplicities $(\overline{[\alpha][\beta]}, \check{F}_i)$ can be obtained with the aid of the Littlewood-Richardson-rule (I, 4.51).

E.g.

$$[2,1][1] = [3,1] + [2^2] + [2,1^2]$$

together with 2.47 and 2.49 yields

$$(\overline{(2,1|1)},F_i) = (\overline{[3,1]} + \overline{[2^2]} + \overline{[2,1^2]}, F_i) = \begin{cases} 2, & i = 1 \\ 3, & i = 2 \end{cases}$$

In this way we obtain for the decomposition matrix of $S_2 \sim S_4$ with respect to $p = 2$:

$$
2.50 \qquad
\begin{matrix} F_1 & F_2 \end{matrix}
$$

$$
\begin{bmatrix}
1 & 0 \\
1 & 1 \\
0 & 1 \\
1 & 1 \\
1 & 0 \\
1 & 0 \\
1 & 1 \\
0 & 1 \\
1 & 1 \\
1 & 0 \\
2 & 1 \\
2 & 3 \\
2 & 1 \\
2 & 2 \\
2 & 2 \\
2 & 2 \\
2 & 2 \\
2 & 1 \\
2 & 3 \\
2 & 1
\end{bmatrix}
\qquad
\begin{matrix}
(4|0) \\
(3,1|0) \\
(2^2|0) \\
(2,1^2|0) \\
(1^4|0) \\
(0|4) \\
(0|3,1) \\
(0|2^2) \\
(0|2,1^2) \\
(0|1^4) \\
(3|1) \\
(2,1|1) \\
(1^3|1) \\
(2|2) \\
(2|1^2) \\
(1^2|2) \\
(1^2|1^2) \\
(1|3) \\
(1|2,1) \\
(1|1^3)
\end{matrix}
$$

With respect to p := 3 we obtain quite similarly:

2.51

$$
\left[\begin{array}{cccccccccccccccc}
1 &&&&&&&&&&&&&&& \\
1 & 1 &&&&&&&&&&&&&& \\
& 1 &&&&&&&&&&&&&& \\
&& 1 &&&&&&&&&&&&& \\
&&& 1 &&&&&&&&&&&& \\
&&&& 1 &&&&&&&&&&& \\
&&&& 1 & 1 &&&&&&&&&& \\
&&&&& 1 &&&&&&&&&& \\
&&&&&& 1 &&&&&&&&& \\
&&&&&&& 1 &&&&&&&& \\
&&&&&&&& 1 &&&&&&& \\
&&&&&&&& 1 & 1 &&&&&& \\
&&&&&&&&& 1 &&&&&& \\
&&&&&&&&&& 1 &&&&& \\
&&&&&&&&&& 1 & 1 &&&& \\
&&&&&&&&&&& 1 &&&& \\
&&&&&&&&&&&& 1 &&& \\
&&&&&&&&&&&&& 1 && \\
&&&&&&&&&&&&&& 1 & \\
&&&&&&&&&&&&&&& 1
\end{array}\right]
\begin{array}{l}
(4|0) \\ (2^2|0) \\ (1^4|0) \\ (3,1|0) \\ (2,1^2|0) \\ (0|4) \\ (0|2^2) \\ (0|1^4) \\ (0|3,1) \\ (0|2,1^2) \\ (3|1) \\ (2,1|1) \\ (1^3|1) \\ (1|3) \\ (1|2,1) \\ (1|1^3) \\ (2|2) \\ (2|1^2) \\ (1^2|2) \\ (1^2|1^2)
\end{array}
$$

(In fact both these matrices can be completed fairly easy in order to obtain the generalized decomposition matrices of $S_2 \sim S_4$ with respect to p = 2,3).

Let us derive the decomposition matrices of $S_2 \sim A_4$ from 2.50, 2.51.

$S_2 \sim A_4$ contains three 2-regular classes, for the conjugacy class of (e; (123)) splits over $S_2 \sim A_4$. Hence one of the two

2-modular irreducible representations F_1, F_2 of $S_2 \sim S_4$ splits over $S_2 \sim A_4$. Since F_1 is one-dimensional, it must be F_2:

$$F_2 \downarrow S_2 \sim A_4 \quad \leftrightarrow \quad F_2^+ + F_2^- .$$

Clifford's theory yields part of the decomposition matrix by cancelling one element of each pair of associated ordinary irreducible representation of $S_2 \sim S_4$. We obtain in this way:

2.52

$F_1 \downarrow S_2 \sim A_4$	F_2^+	F_2^-	
1	0	0	$(4\|0) \downarrow S_2 \sim A_4$
1	1	1	$(3,1\|0) \downarrow S_2 \sim A_4$
1	0	0	$(0\|4) \downarrow S_2 \sim A_4$
1	1	1	$(0\|3,1) \downarrow S_2 \sim A_4$
2	1	1	$(3\|1) \downarrow S_2 \sim A_4$
2	2	2	$(2\|2) \downarrow S_2 \sim A_4$
2	2	2	$(2\|1^2) \downarrow S_2 \sim A_4$
2	1	1	$(1\|3) \downarrow S_2 \sim A_4$

It remains to evaluate the decomposition numbers of the irreducible constituents $(\alpha|\beta)^{\pm}$ of

$$(2^2|0) \downarrow S_2 \sim A_4 \quad = \quad (2^{2^+}|0) + (2^{2^-}|0)$$
$$(0|2^2) \downarrow S_2 \sim A_4 \quad = \quad (0|2^{2+}) + (0|2^{2-})$$
$$(2,1|1) \downarrow S_2 \sim A_4 \quad = \quad (2,1^+|1 \quad + (2,1^-|1)$$
$$(1|2,1) \downarrow S_2 \sim A_4 \quad = \quad (1|2,1^+) + (1|2,1^-) .$$

Since $F_2 = \overline{(2^2|0)}$, $\overline{(2^{2\pm}|0)}$ are irreducible. We may number the constituents of $F_2 \downarrow S_2 \sim A_4$ so that

$$(2^{2+}|0) = F_2^+ .$$

Using 2.52 we are able to evaluate the matrix of Brauer characters of $S_2 \sim A_4$, it is the matrix

2.53
$$\begin{array}{ccc} \begin{pmatrix} 4&0&0&0 \\ 0&0&0&0 \end{pmatrix} & \begin{pmatrix} 1&0&1&0 \\ 0&0&0&0 \end{pmatrix}^+ & \begin{pmatrix} 1&0&1&0 \\ 0&0&0&0 \end{pmatrix}^- \end{array}$$
$$\begin{bmatrix} 1 & 1 & 1 \\ 1 & \frac{i\sqrt{3}-1}{2} & \frac{1+i\sqrt{3}}{2} \\ 1 & \frac{1+i\sqrt{3}}{2} & \frac{i\sqrt{3}-1}{2} \end{bmatrix}$$

Using this, some characters of $S_2 \sim S_4$ and $S_2 \sim A_4$ as well as 2.50 we obtain the decomposition matrix of $S_2 \sim A_4$ with respect to $p = 2$. It is the matrix

2.54

$F_1 \downarrow S_2 \wr A_4$	F_2^+	F_2^-	
1	0	0	$(4\|0) \downarrow S_2 \wr A_4$
1	1	1	$(3,1\|0) \downarrow S_2 \wr A_4$
0	1	0	$(2^{2+}\|0)$
0	0	1	$(2^{2-}\|0)$
1	0	0	$(0\|4) \downarrow S_2 \wr A_4$
1	1	1	$(0\|3,1) \downarrow S_2 \wr A_4$
0	1	0	$(0\|2^{2+})$
0	0	1	$(0\|2^{2-})$
2	1	1	$(3\|1) \downarrow S_2 \wr A_4$
1	2	1	$(2,1^+\|1)$
1	1	2	$(2,1^-\|1)$
2	2	2	$(2\|2) \downarrow S_2 \wr A_4$
2	2	2	$(2\|1^2) \downarrow S_2 \wr A_4$
2	1	1	$(1\|3) \downarrow S_2 \wr A_4$
1	2	1	$(1\|2,1^+)$
1	1	2	$(1\|2,1^-)$

In order to evaluate the decompostion matrix of $S_2 \wr S_{4A_2}$ with respect to p = 2, we notice first that $S_2 \wr S_{4A_2}$ possesses exactly two 2-modular irreducible representation since no 2-regular class of $S_2 \wr S_4$ splits over $S_2 \wr S_{4A_2}$. Hence both $F_1 \downarrow S_2 \wr S_{4A_2}$ and $F_2 \downarrow S_2 \wr S_{4A_2}$ are irreducible. We thus obtain the decomposition matrix directly from 2.50. It is the matrix

2.55

$F_1 \downarrow S_2 \sim S_{4A_2}$	$F_2 \downarrow S_2 \sim S_{4A_2}$	
1	0	$(4\mid 0) \downarrow S_2 \sim S_{4A_2}$
1	1	$(3,1\mid 0) \downarrow S_2 \sim S_{4A_2}$
0	1	$(2^2\mid 0) \downarrow S_2 \sim S_{4A_2}$
1	1	$(2,1^2\mid 0) \downarrow S_2 \sim S_{4A_2}$
1	0	$(1^4\mid 0) \downarrow S_2 \sim S_{4A_2}$
2	1	$(3\mid 1) \downarrow S_2 \sim S_{4A_2}$
2	3	$(2,1\mid 1) \downarrow S_2 \sim S_{4A_2}$
2	1	$(1^3\mid 1) \downarrow S_2 \sim S_{4A_2}$
1	1	$(2\mid 2)_+$
1	1	$(2\mid 2)_-$
1	1	$(2\mid 1^2) \downarrow S_2 \sim S_{4A_2}$
1	1	$(1^2\mid 2) \downarrow S_2 \sim S_{4A_2}$
1	1	$(1^2\mid 1^2)_+$
1	1	$(1^2\mid 1^2)_-$

The decomposition matrix of $S_2 \sim S_{4A_2}^{A_4}$ with respect to $p = 2$ turns out to be

2.56

	$F_1 \downarrow S_2 \wr S_{4A_2}^{A_4}$	$F_2 \downarrow S_2 \wr S_{4A_2}^{A_4}$	
	1	0	$(4\|0) \downarrow S_2 \wr S_{4A_2}^{A_4}$
	1	1	$(3,1\|0) \downarrow S_2 \wr S_{4A_2}^{A_4}$
	0	1	$(2^2\|0) \downarrow S_2 \wr S_{4A_2}^{A_4}$
	1	1	$(2,1^2\|0) \downarrow S_2 \wr S_{4A_2}^{A_4}$
	1	0	$(1^4\|0) \downarrow S_2 \wr S_{4A_2}^{A_4}$
	2	1	$(3\|1) \downarrow S_2 \wr S_{4A_2}^{A_4}$
	2	3	$(2,1\|1) \downarrow S_2 \wr S_{4A_2}^{A_4}$
	2	1	$(1^3\|1) \downarrow S_2 \wr S_{4A_2}^{A_4}$
	2	2	$(2\|2) \downarrow S_2 \wr S_{4A_2}^{A_4}$
	1	1	$(2\|1^2)_+$
	1	1	$(2\|1^2)_-$
	2	2	$(1^2\|1^2) \downarrow S_2 \wr S_{4A_2}^{A_4}$
	1	1	$(1^2\|2)_+$
	1	1	$(1^2\|2)_-$

This done one may ask for the generalized decomposition numbers which complete the decomposition matrix (see part I). Numerical results concerning can be found in Celik/Pahlings/Kerber [1].

There is also a theorem concerning the question when the generalized decomposition numbers are rational integral. Sufficient for this is that the values of the ordinary characters on p-singular elements are rational integral (Reynolds [1]). This together with 2.18 yields

2.57 The decomposition numbers of $G \sim H$ are rational integral, if the character tables of G as well as of the inertia factors $H \cap S_{(n)}$ have rational integral entries only. This holds in particular for $G \sim S_n$ if the character table of G is rational integral.

2.46 yields the following corollary:

2.58 The generalized decomposition numbers of each member of the series of Weyl groups are rational integral.

With this we have shown some applications of the construction of irreducible representations of wreath products as it is described in 2.15.

2.15 says that the irreducible representations of $G \sim H$ are of the form

$$F = (\widetilde{F^*} \otimes F') \uparrow G \sim H,$$

where $\widetilde{F^*}$ denotes the extension of F^* of G^* to its inertia group $G \sim H_{F^*}$ as it is described by 2.9. In the case when all the factors of F^* are equal: say to the representation D of G, then $\widetilde{F^*}$ is just the representation

$$(D; IH),$$

where IH denotes the identity representation of H. In the same way as the irreducible representation $\overset{n}{\#}D$ can be extended to $\widetilde{\overset{n}{\#}D} = (D; IH)$, we can extend $\overset{n}{\#}(D_1+D_2)$, where D_1, D_2 are arbitrary (i.e. may be reducible) representations of G to $G \sim H$.

Hence the question arises for the decomposition of $(D_1+D_2; IH)$. It will later turn out to be useful to know the following result of F. Sänger (hitherto unpublished):

<u>2.59</u> <u>If D_1 and D_2 are representations of a finite group G (which may be reducible) over a field K, then</u>

$$\underline{\widetilde{\overset{n}{\#}(D_1+D_2)} = \sum_{k=0}^{n} (\widetilde{\overset{k}{\#} D_1} \# \widetilde{\overset{n-k}{\#} D_2}) \uparrow G \sim S_n.}$$

Proof: Let D_i be afforded by the module M_i with underlying vector space V_i, $i = 1,2$.

Then $\widetilde{\#^n(D_1+D_2)}$ is afforded by the left $KG\backslash S_n$-module $\widetilde{\#^n(M_1 \dot{+} M_2)}$ with underlying vector space

$$W := \overset{n}{\otimes}(V_1 \dot{+} V_2) = \bigoplus_{k=0}^{n} W_k,$$

where each W_k is a left $KG\backslash S_n$-module and

$$\begin{aligned} W_k = {} & \underbrace{V_1 \otimes \ldots \otimes V_1}_{k \text{ summands}} \otimes \underbrace{V_2 \otimes \ldots \otimes V_2}_{n-k} \\ & \oplus \underbrace{V_1 \otimes \ldots \otimes V_1}_{k-1} \otimes V_2 \otimes V_1 \otimes \underbrace{V_2 \otimes \ldots \otimes V_2}_{n-k-1} \\ & \oplus \underbrace{V_1 \otimes \ldots \otimes V_1}_{k-2} \otimes V_2 \otimes V_1 \otimes V_1 \otimes \underbrace{V_2 \otimes \ldots \otimes V_2}_{n-k-1} \\ & \cdot \\ & \cdot \\ & \cdot \\ & \oplus \underbrace{V_2 \otimes \ldots \otimes V_2}_{n-k} \otimes \underbrace{V_1 \otimes \ldots \otimes V_1}_{k} \\ = {} & \bigoplus_{\sigma \in L} (e;\sigma) (\underbrace{V_1 \otimes \ldots \otimes V_1}_{k} \otimes \underbrace{V_2 \otimes \ldots \otimes V_2}_{n-k}) \end{aligned}$$

where L is a complete system of representatives of the left cosets of $S_k \times S_{n-k}$ in S_n. Hence Lemma (44.1) in Curtis/Reiner [1] yields that W_k affords the representation of $G\sim S_n$ which is induced by the representation afforded by the module with underlying vector space

$$V_1 \otimes \ldots \otimes V_1 \otimes V_2 \otimes \ldots \otimes V_2.$$

But the module with this underlying vector space affords the representation

$$\underbrace{\underbrace{\overbrace{\# D_1}^{k}}_{\text{of } G\sim S_k} \;\#\; \underbrace{\overbrace{\# D_2}^{n-k}}_{\text{of } G\sim S_{n-k}}}_{\text{of } G\sim(S_k \times S_{n-k})}$$

q.e.d.

An example is provided by

$$\overbrace{\#([2]+[1^2])}^{3} = \overbrace{\#[2]}^{3} + (\overbrace{[2]\#[2]\#[1^2]} \otimes ([2]\#[1])')\uparrow S_2\sim S_3$$
$$+ (\overbrace{[2]\#[1^2]\#[1^2]} \otimes ([1]\#[2])')\uparrow S_2\sim S_3 + \overbrace{\#[1^2]}^{3}$$
$$= (2;3) + \overbrace{\#[2]\#[1^2]}^{2}\uparrow S_2\sim S_3 + \overbrace{[2]\#[1^2]}^{2}\uparrow S_2\sim S_3 + (1^2;\ 3).$$

Having obtained a result on the decomposition of the extension of the reducible representation $\overset{n}{\#}(D_1+D_2)$ to $G\sim S_n$ let us conclude this section with a hint to another way of producing representations of $G\sim H$ from representations D of G.

The method described above arose from the desire to apply Clifford's theory of representations of groups with normal subgroups where one has to start from irreducible representations of the normal subgroup in question. Hence we were forced to begin with irreducible representations of the basis group which are just the outer tensor products $F^* = \# F_i$ of irreducible representations F_i of G (if we assume the groundfield to be algebraically closed).

If we are not forced to obtain irreducible representations of $G\sim H$ we may start from reducible representations of G^* as well and there is in fact a way to do that and apply an extension process quite similar to 2.3 which yields an in general reducible representation of $G\sim H$ which will turn out to be useful later on.

Let F_G denote a representation of a group G over a field K and let V denote the representation space, M the representation module. If n is a given natural number, then we may form the n-fold outer direct sum $\tilde{+}M$ of M with itself. The underlying vector space is $\overset{n}{\times}V$ and the operation of $G^* \leq G\sim H$ $(H \leq S_n)$:

2.60 $$(f;\, 1_H)\,(v_1,\ldots,v_n) := (f(1)v_1,\ldots,f(n)v_n)$$

It can be extended to $G\sim H$ in a fashion quite similar to 2.3:

2.61 $(f;\pi)(v_1,\ldots,v_n) := (f(1)v_{\pi^{-1}(1)},\ldots,f(n)\,v_{\pi^{-1}(n)})$.

We denote this module by $\widetilde{\dot{+}^{n}}\,M$, the afforded representation by $\widetilde{\dot{+}^{n}}\,F_G$.

It is easy to check that the character values of $\widetilde{\dot{+}^{n}}\,F_G$ are as follows

2.62 $$\chi^{\widetilde{\dot{+}^{n}}F_G}(f;\pi) = \sum_{\substack{i\\ \pi(i)=i}} \chi^{F_G}(f(i)).$$

If e.g. ν is the natural representation of $G \leq S_m$, then it is not difficult to see that $\widetilde{\dot{+}^{n}}\,\nu$ induces on the natural basis $\{(0,\ldots,0,e_j,0,\ldots,0) \mid 1 \leq e_j \leq m\}$ (where ν acts on $\mathbb{C}^m = \langle\langle e_1,\ldots,e_m\rangle\rangle$ by $\nu(g)\colon e_i \mapsto e_{g(i)}$) a permutation group similar to $\varphi[G\wr H]$, where φ denotes the permutation representation 1.4.

Chapter II

An application to representation theory:

Symmetrization of inner tensor products of representations

The results of the preceding chapter are applied to the theory of symmetrizing the n-fold inner tensor power $\overset{n}{\otimes} F_G$ of an ordinary representation F_G of a group G with ordinary irreducible representations $[\alpha]$ of S_n.

Some applications are discussed and the case $G := S_m$ is considered in more detail.

3. Symmetrized inner products of representations

Let F_G denote a linear representation of a group G over a field K with representation module M and underlying vector space V and a corresponding matrix representation $\mathbb{F}_G$.

In section 2 we have seen, how an additionally given natural number n leads to a left G^*-module $\overset{n}{\#} M$ with underlying vector space $\overset{n}{\otimes} V$ which affords the representation $\overset{n}{\#} F_G$ of $G^* = G_1 \times \ldots \times G_n \leq G \wr S_n$, and how a corresponding matrix representation $\overset{n}{\#} \mathbb{F}_G$ can be defined. And we have seen in section 2, how this representation $\overset{n}{\#} F_G$ can be extended to a representation $\widetilde{\overset{n}{\#}} F_G$ of $G \wr S_n$, which is of the same dimension as $\overset{n}{\#} F_G$, a corresponding matrix representation was denoted by $\widetilde{\overset{n}{\#}} \mathbb{F}_G$.

In this section I would like to show, that it can be very useful to consider, for a given F_G, suitable natural numbers and restrictions of the corresponding representations $\overset{n}{\#} F_G$ or $\widetilde{\overset{n}{\#}} F_G$ to certain subgroups of $G \wr S_n$ as well as to induce to suitable groups which contain G^* or even $G \wr S_n$ as subgroups.

An interesting example is provided by a proof given by Tate (cf. Serre [1]) of the well known theorem, that the dimension f^{F_G} of each ordinary irreducible representation F_G of a given finite group G divides the index $|G: C(G)|$ of the centre $C(G)$ of G.

In order to prove this, we form, for a given natural number $n \in \mathbb{N}$, the representation $\overset{n}{\#} F_G$ of $G^* \leq G \sim S_n$, which has $\overset{n}{\otimes} V$ as a representation space, if V denotes the representation space of F_G.

A subgroup of G^*, which has not been mentioned in section 1, is

$$U := \{(f;1) \mid f: \mathbb{N}_n \to C(G) \wedge \prod_{i=1}^{n} f(i) = 1_G\}$$

$$= \{(g_1,\ldots,g_n;\ 1) \mid g_i \in C(G) \wedge \prod_{i=1}^{n} g_i = 1_G\}.$$

The order of this subgroup is

$$|U| = |C(G)|^{n-1} .$$

The elements of U act on the generating elements $v_1 \otimes \ldots \otimes v_n$ of $\overset{n}{\otimes} V$, $v_i \in V$, as follows:

$$(g_1,\ldots,g_n;\ 1)\ v_1 \otimes \ldots \otimes v_n = g_1 v_1 \otimes \ldots \otimes g_n v_n.$$

Since $g_i \in C(G)$, $g_i v_i = \varkappa_i v_i$, $\varkappa_i \in \mathbb{C}$, and since $\prod g_i = 1_G$, $\prod \varkappa_i = 1_{\mathbb{C}}$, so that the irreducible representation $\overset{n}{\#} F_G$ of U acts trivially on $\overset{n}{\otimes} V$:

$\forall\ (f;1) \in U,\ v_1 \otimes \ldots \otimes v_n \in \overset{n}{\otimes} V,\ ((f;1)\ v_1 \otimes \ldots \otimes v_n = v_1 \otimes \ldots \otimes v_n).$

Hence U is contained in the kernel of $\overset{n}{\#} F_G$ so that we obtain an ordinary irreducible representation F of G^*/U by putting

$$F((f;1)U) := \overset{n}{\#} F_G(f;1).$$

The dimension of F is $(f^{F_G})^n$, and it divides the order $|G|^n/|C(G)|^{n-1}$ of the represented group, i.e. we have

$$\forall\ n\in\mathbb{N}\left(\frac{|G : C(G)|^n}{(f^{F_G})^n} \in \frac{1}{|C(G)|}\cdot\mathbb{Z}\right),$$

which shows, that $|G : C(G)|/f^{F_G}$ is a natural number.

q.e.d.

An example using $\overset{n}{\#} F_G$ and induction was discussed in volume I.

We considered a finite permutation group G, say $G \leq S_m$, and a representation F_G of G, together with a second permutation group $H \leq S_n$ with a representation F_H, F_G and F_H over a field K. We formed $\widetilde{\overset{n}{\#} F_G}$ of $G\sim H$ and F_H', defined by

$$\forall\ (f;\pi) \in G\sim H\ (F_H'(f;\pi) := F_H(\pi)).$$

Since 1.4 yields an embedding of $G\sim H$ into S_{mn}, the inner tensor product $(F_G;F_H) := \widetilde{\overset{n}{\#} F_G} \otimes F_H'$ allows to define the following representation:

<u>3.1</u> $$F_G \odot F_H := \underbrace{(F_G;F_H)}_{\text{of } \varphi[G\sim H]} \uparrow S_{mn} = \widetilde{\overset{n}{\#} F_G} \otimes F_H' \uparrow S_{mn}.$$

This in general reducible representation of S_{mn} was called <u>the symmetrized outer product of</u> F_G <u>and</u> F_H (cf. I §§ 5,6).

If on the other hand we restrict $\widetilde{\overset{n}{\#} F_G}$ to the diagonal of G^*, we obtain the following interesting representation of G:

3.2 $$\overset{n}{\otimes} F_G := \overset{n}{\#} F_G \downarrow \text{diag } G^* = \widetilde{\overset{n}{\#} F_G} \downarrow \text{diag } G^*,$$

which is in general reducible.

Notice that

3.3 $$\forall\, g \in G\ (\overset{n}{\otimes} F_G(g) = \widetilde{\overset{n}{\#} F_G}(g,\ldots,g;1_H)).$$

The question arises which are the irreducible constituents of $\overset{n}{\otimes} F_G$.

This problem can be attacked (but not always solved) with the aid of the representation theory of permutation groups H of degree n, e.g. $H := S_n$. A special case of this is in fact the famous discovery of Schur, van der Waerden and Weyl, that there is a close connection between the representation theories of general linear and symmetric groups (cf. Schur [1], [2], van der Waerden [1], Weyl [1], [2], [3], cf. also Curtis/ Reiner [1], § 67, Kerber [8]). They considered the identity mapping on $GL(n,\mathbb{C})$ as an ordinary representation of $GL(n,\mathbb{C}) : F_G := id_{GL(n,\mathbb{C})}$, and obtained important results by using the ordinary representation theory of S_n.

Hence let us assume, that G is a group, n a natural number, $H \leq S_n$, K a field and F_G to be a linear representation of G over K.

We notice first, that the composition law of $G \wr H$ implies, that the elements of diag G^* commute with the elements of H':

3.4 $\forall\, g\in G, \pi\in H\ ((g,\ldots,g;\pi)=(g,\ldots,g;1_H)(e;\pi)=(e;\pi)(g,\ldots,g;1_H))$,

i.e. diag $G^* \leq C_{G\sim H}(H') \wedge H' \leq C_{G\sim H}(\mathrm{diag}\ G^*)$.

Hence the corresponding elements of the image of this representation commute as well:

3.5 $\forall\, g \in G, \pi \in H\ (\widetilde{\#F_G^{n}}(g,\ldots,g;\pi) = \widetilde{\#F_G^{n}}(g,\ldots,g;1_H)\widetilde{\#F_G^{n}}(e;\pi)$

$= \widetilde{\#F_G^{n}}(e;\pi)\widetilde{\#F_G^{n}}(g,\ldots,g;1_H)$.

Hence by putting

3.6 (i) $\forall\, g \in G\ (\hat{F}_G(g) := \widetilde{\#F_G^{n}}(g,\ldots,g;1_H))$,

(ii) $\forall\, \pi \in H\ (\check{F}_G(\pi) := \widetilde{\#F_G^{n}}(e;\pi))$,

we obtain from a given representation F_G of G and a natural number $n \in \mathbb{N}$ a representation $\hat{F}_G$ of G and a representation $\check{F}_G$ of H so that the elements of their images commute:

3.7 $\forall\, g \in G, \pi \in H\ (\widetilde{\#F_G^{n}}(g,\ldots,g;\pi) = \hat{F}_G(g)\check{F}_G(\pi) = \check{F}_G(\pi)\hat{F}_G(g))$.

Notice that the following holds:

3.8 $$\hat{F}_G = \overset{n}{\otimes} F_G.$$

The crucial fact is 3.7. It shows, that we can apply a corollary of Schur's lemma (cf. Boerner [2], I § 8), in the case when the groundfield K is algebraically closed and $\mathrm{char}K \nmid |H|$. Let us assume that this is valid. It implies that $\check{F}_G$ is completely reducible, so that we can choose a basis of the representation space $\overset{n}{\otimes} V$ of $\check{F}_G$ which yields a corresponding matrix representation $\check{\mathbb{F}}_G$ in its completely reduced form, say

$$\underline{\underline{3.9}} \quad \forall\ \pi\in H \quad (\check{\mathbb{F}}_G(\pi) = \widetilde{{}^n\mathbb{F}_G}(e;\pi) = \begin{bmatrix} \ddots \mathbb{D}^i(\pi) & & & 0 \\ & \ddots & \mathbb{D}^i(\pi) \Big\} n_i\text{-times} & \\ 0 & & & \ddots \end{bmatrix}$$

$$= \dot{+}_{i=1}^{t} n_i \mathbb{D}^i(\pi) = \dot{+}_{\substack{i \\ n_i>0}} (\mathbb{I}_{n_i} \times \mathbb{D}^i(\pi)),$$

where $D^1,\dots,D^t$ denotes a complete system of pairwise inequivalent and irreducible K-representations of H and where $\mathbb{I}_{n_i}$ is the n_i-rowed identity matrix.

The corollary of Schur's lemma now implies that the same basis of $\overset{n}{\otimes} V$ yields a corresponding decomposition of $\hat{F}_G$. If f^i denotes the dimension of D^i, then there are matrices $\mathbb{F}_G \boxdot \mathbb{D}^i(g)$, which satisfy the following:

$$\underline{\underline{3.10}} \quad \forall\ g \in G \quad (\hat{\mathbb{F}}_G(g) = \dot{+}_{\substack{i \\ n_i>0}} ((\mathbb{F}_G \boxdot \mathbb{D}^i)(g) \times \mathbb{I}_{f^i}),$$

These matrices $\mathbb{F}_G \boxdot \mathbb{D}^i(g)$ form a matrix representation of G.

The corresponding representation $F_G \boxdot D^i$ of G hence satisfies:

3.11 If K is algebraically closed and $\mathrm{char}K \nmid |H|$, then to each irreducible constituent D^i of $\check{F}_G$ there corresponds a certain (and in general reducible) constituent $F_G \boxdot D^i$ of $\hat{F}_G$ which we call the symmetrized inner product of F_G and D^i, following Robinson's notation (Robinson [5]). It satisfies

$$\forall\, g \in G, \pi \in H \ (\widetilde{\overset{n}{\#}F_G}(g,\dots,g;\pi) = \hat{F}_G(g)\check{F}_G(\pi)$$

$$= \dot{+}_{\substack{i \\ n_i>0}} (F_G \boxdot D^i(g) \times D^i(\pi))).$$

The dimension of $F_G \boxdot D^i$ is the multiplicity n_i of D^i in $\check{F}_G$ and it occurs f^i-times in $\hat{F}_G$, if f^i denotes the dimension of D^i, so that we obtain

$$\hat{F}_G = \sum_{\substack{i \\ n_i>0}} f^i\, (F_G \boxdot D^i).$$

This yields (apply 3.8):

$$\overset{n}{\otimes} F_G = \sum_{\substack{i \\ n_i>0}} f^i\, (F_G \boxdot D^i).$$

Before specializing on G and H let us continue this section with some preliminary remarks concerning corresponding characters to become a bit more acquainted with this useful concept of symmetrization of representations.

One of the main questions which arise is the question, which irreducible representations D^i of H occur in $\check{F}_G$. If $\mathrm{char}K = 0$, we may use characters and their orthogonality relations, which yield

$$3.12 \qquad n_i = (\check{F}_G, D^i) = \frac{1}{|H|} \sum_{\pi \in H} \chi^{\check{F}_G}(\pi)\zeta^{D^i}(\pi^{-1}).$$

In order to evaluate $\chi^{\check{F}_G}(\pi)$, we use 2.7, which gives

$$\underline{3.13} \quad \underline{\forall\, g \in G, \pi \in H \; (\chi^{\widetilde{n}\#F_G}(g,\ldots,g;\pi) = \prod_{k=1}^{n} \chi^{F_G}(g^k)^{a_k(\pi)}),}$$

so that especially

$$\underline{3.14} \quad \underline{\forall\, \pi \in H \; (\chi^{\check{F}_G}(\pi) = \chi^{\widetilde{n}\#F_G}(e;\pi) = (f^{F_G})^{c(\pi)}).}$$

Hence 3.12 implies

$$\underline{3.15} \quad \underline{(\check{F}_G, D^i) = \frac{1}{|H|} \sum_{\pi \in H} (f^{F_G})^{c(\pi)}\zeta^{D^i}(\pi).}$$

(Since G has ordinary representations of dimension m for every natural number m, we obtain as a byproduct the maybe surprising corollary that for a subgroup H of S_n and $m \in \mathbb{N}$, the number

$$\frac{1}{|H|}\sum_{\pi \in H} m^{c(\pi)}\zeta(\pi)$$

is a nonnegative integer for any ordinary irreducible character ζ of H.)

3.15 gives us

<u>3.16</u> <u>If F_G is an ordinary representation of a group G and D an ordinary irreducible representation of a permutation group H of finite degree, then the symmetrized inner product $F_G \boxdot D$ exists if and only if</u>

$$\sum_{\pi \in H} (f^{F_G})^{c(\pi)}\zeta^D(\pi) \neq 0.$$

Hence for example $F_G \boxdot IH$ always exists. For its dimension we obtain

<u>3.17</u> $$f^{F_G \boxdot IH} = (\check{F}_G, IH) = \frac{1}{|H|}\sum_{\pi \in H}(f^{F_G})^{c(\pi)}.$$

Taking $H := S_n$ we get

<u>3.18</u> $$f^{F_G \boxdot [n]} = \frac{1}{n!}\sum_{\pi \in S_n}(f^{F_G})^{c(\pi)}.$$

Analogously we obtain for the alternating representation $[1^n]$ of S_n:

3.19 <u>If $F_G \boxdot [1^n]$ exists, then it is of dimension</u>

$$f^{F_G \boxdot [1^n]} = \frac{1}{n!} \sum_{\pi \in S_n} \epsilon_\pi (f^{F_G})^{c(\pi)} .$$

For example

$$f^{F_G \boxdot [1^2]} = \frac{1}{2} ((f^{F_G})^2 - f^{F_G}) = \binom{f^{F_G}}{2}.$$

In fact one can show that

3.20 $$f^{F_G \boxdot [1^n]} = \binom{f^{F_G}}{n}, \text{ if } F_G \boxdot [1^n] \text{ exists,}$$

and that $F_G \boxdot [1^n]$ exists if and only if $n \leq f^{F_G}$. We shall return to this question of existence of $F_G \boxdot D^i$ later on. Then we shall also discuss, when $F_G \boxdot D^i$ is irreducible.

3.11 allows an evaluation of the character of $F_G \boxdot D^i$. We need only apply the orthogonality relations to the equation

3.21 $$\chi^{\widetilde{n \# F_G}}(g,\ldots,g;\pi) = \sum_{\substack{i \\ n_i > 0}} \chi^{F_G \boxdot D^i}(g) \zeta^{D^i}(\pi),$$

which is an immediate consequence of 3.11.

We multiply both sides by $\zeta^{D^j}(\pi^{-1})$, where D^j denotes an irreducible K-representation of H. By summing up over all the elements π of H , we obtain from the orthogonality relations and 2.7 (i):

3.22 If K is algebraically closed and $\operatorname{char}K \nmid |H|$, then if $F_G \boxdot D^i$ exists, its character reads as follows:

$$\forall\, g \in G\ \left(\chi^{F_G \boxdot D^i}(g) = \frac{1}{|H|} \sum_{\pi \in H} \zeta^{D^i}(\pi^{-1}) \prod_{k=1}^{n} \chi^{F_G}(g^k)^{a_k(\pi)}\right).$$

If $F_G \boxdot D^i$ does not exist, then we have

$$\forall\, g \in G\ \left(\frac{1}{|H|} \sum_{\pi \in H} \zeta^{D^i}(\pi^{-1}) \prod_{k=1}^{n} \chi^{F_G}(g^k)^{a_k(\pi)} = 0\right).$$

Insertion into $\overset{n}{\otimes} F_G = \Sigma f^{D^i}(F_G \boxdot D^i)$ yields

3.23 $$\forall\, g \in G\ \left(\chi^{F_G}(g)^n = \frac{1}{|H|} \sum_{\substack{i \\ n_i > 0}} f^{D^i} \sum_{\pi \in H} \zeta^{D^i}(\pi^{-1}) \prod_{k=1}^{n} \chi^{F_G}(g^k)^{a_k(\pi)}\right).$$

Let us again consider the special case $H := S_n$, $K := \mathbb{C}$. Then we obtain for an ordinary representation F_G of G (of sufficiently high dimension) and each $g \in G$:

3.24 (i) $\chi^{F_G \boxdot [1]}(g) = \chi^{F_G}(g)$,

(ii) $\chi^{F_G \boxdot [2]}(g) = \frac{1}{2}\left(\chi^{F_G}(g)^2 + \chi^{F_G}(g^2)\right)$,

(iii) $\chi^{F_G \boxdot [1^2]}(g) = \frac{1}{2}\left(\chi^{F_G}(g)^2 - \chi^{F_G}(g^2)\right)$,

(iv) $\chi^{F_G \boxdot [3]}(g) = \frac{1}{6}\left(\chi^{F_G}(g)^3 + 3\chi^{F_G}(g)\chi^{F_G}(g^2) + 2\chi^{F_G}(g^3)\right)$,

(v) $\chi^{F_G \boxdot [2,1]}(g) = \frac{1}{6}\left(2\chi^{F_G}(g)^3 - 2\chi^{F_G}(g^3)\right)$,

(vi) $\underline{\chi^{F_G \boxdot [1^3]}(g) = \frac{1}{6}\,(\chi^{F_G}(g)^3 - 3\chi^{F_G}(g)\chi^{F_G}(g^2) + 2\chi^{F_G}(g^3)),}$

<u>and so on (use the character tables of symmetric groups).</u>

3.24 provides a useful method for the evaluation of character tables. It is well known, that if F_G is a faithful representation of G, G a finite group, then every irreducible K-representation of G occurs under the irreducible constituents of $\overset{n}{\otimes} F_G$, $n = 1,2,\ldots,t$, t the number of different values of χ^{F_G}. Thus if we are given such a faithful representation F_G, the orders of the conjugacy classes of G as well as the class in which g^n lies, for each $g \in G$ and some $n \in \mathbb{N}$, then we may evaluate the characters of the tensor power $\overset{n}{\otimes} F_G$ and use 3.11, 3.22, 3.24 to break them up by forming the characters of symmetrized inner products $F_G \boxdot [\alpha]$, $\alpha \vdash n$. The method is to subtract known irreducible characters of G if they are contained in $\chi^{F_G \boxdot [\alpha]}$ and check, whether the remaining character of G is irreducible or not.

It is often possible to seperate further irreducible characters of G in this way, say by a man-machine interaction program (cf. Esper [1]).

These results have been obtained by specializing on H. If on the other hand we specialize on G we obtain famous results concerning the connection between the ordinary irreducible representations of $H := S_n$ and $G := GL(m,\mathbb{C})$ if we put

$$F_{GL(m,\mathbb{C})} := id_{GL(m,\mathbb{C})} : GL(m,\mathbb{C}) \to GL(m,\mathbb{C}) : g \mapsto g.$$

For in this case it turns out, that $F_G \boxdot [\alpha] = id_{GL(m,\mathbb{C})} \boxdot [\alpha]$ is irreducible if it exists. We shall return to this later on.

I would like to continue this section with some of the results of van Zanten and de Vries (van Zanten/de Vries [1]). They have pointed out, how theorems concerning the number of solutions of the equation

3.25 $$x^n = g$$

for a given element g of a finite group G and a given natural number n can be derived using the above results on the partial reduction of $\overset{n}{\otimes} F_G$.

3.26 Let G denote a finite group, g a fixed element of G and n a natural number. If

$$||\{x \mid x \in G \wedge x^n = g\}|| \neq 1,$$

then there is at least one ordinary irreducible representation F_G of G which satisfies $F_G \neq IG$ and so that $\overset{n}{\otimes} F_G$ contains IG at least once as an irreducible constituent.

Proof: We assume on the contrary, that each ordinary irreducible representation F_G different from the identity representation IG of G is such that $\overset{n}{\otimes} F_G$ does not contain IG, i.e. we assume that

$$\forall F_G \neq IG \left(\frac{1}{|G|} \sum_{x \in G} \chi^{F_G}(x)^n = 0\right) \tag{1}$$

An application of 3.11 yields (take $H := S_n$):

$$\forall F_G \neq IG \left(\frac{1}{|G|} \sum_{\substack{\alpha \vdash n \\ n_\alpha > 0}} f^\alpha \sum_{x \in G} \chi^{F_G \boxdot [\alpha]}(x) = 0\right),$$

so that we obtain for each partition $\alpha \vdash n$ with $n_\alpha = (\check{F}_G, [\alpha]) > 0$:

$$(F_G \neq IG \wedge n_\alpha > 0) \Rightarrow \frac{1}{|G|} \sum_{x \in G} \chi^{F_G \boxdot [\alpha]}(x) = 0. \tag{2}$$

Let us compare these results with the character value

$$\begin{aligned} \chi^{\widetilde{\overset{n}{\#} F_G}}(x,\ldots,x;(1\ldots n)) &= \chi^{F_G}(x^n) \\ &= \sum_{\substack{\alpha \vdash n \\ n_\alpha > 0}} \chi^{F_G \boxdot [\alpha]}(x)\zeta^\alpha((1\ldots n)) \end{aligned} \tag{3}$$

(use 3.13 and 3.21). (3) yields

$$\frac{1}{|G|} \sum_{x \in G} \chi^{F_G}(x^n) = \sum_{\substack{\alpha \vdash n \\ n_\alpha > 0}} \zeta^\alpha((1\ldots n)) \frac{1}{|G|} \sum_{x \in G} \chi^{F_G \boxdot [\alpha]}(x),$$

so that an application of (2) yields

$$\forall F_G \neq IG \left(\frac{1}{|G|} \sum_{x \in G} \chi^{F_G}(x^n) = 0\right). \tag{4}$$

If now $r(n,g)$ denotes the number of n-th roots x of

$g \in G$, then it satisfies the equation

$$\sum_{x \in G} \chi^{F_G}(x^n) = \sum_{g \in G} r(n,g)\chi^{F_G}(g),$$

so that by the orthogonality relations we obtain

$$r(n,g) = \frac{1}{|G|}\sum_{F_G}\sum_{x \in G} \chi^{F_G}(x^n)\chi^{F_G}(g^{-1}), \qquad (5)$$

where the first summation is over all the ordinary irreducible representations F_G of G.

Equation (4) now shows, that IG yields the only contribution to this sum, so that we get

$$r(n,g) = \frac{1}{|G|}\cdot|G| = 1,$$

which contradicts our assumption.

q.e.d.

Remark: to prove this we used the following

<u>3.27</u> <u>If G is a finite group with an ordinary irreducible representation F_G, then for each $n \in \mathbb{N}$, $g \in G$:</u>

(i) $$\chi^{F_G}(g^n) = \sum_{\substack{\alpha \vdash n \\ n_\alpha > 0}} \zeta^\alpha((1\ldots n))\chi^{F_G \boxdot [\alpha]}(g),$$

(ii) $$\chi^{F_G}(g)^n = \sum_{\substack{\alpha \vdash n \\ n_\alpha > 0}} f^\alpha \chi^{F_G \boxdot [\alpha]}(g),$$

where $n_\alpha := (\check{F}_G, [\alpha]) = \frac{1}{n!}\sum_{\pi \in S_n} \zeta^\alpha(\pi)(f^{F_G})^{c(\pi)}$.

It seems remarkable to notice that this implies

3.28 If G is a finite group with an ordinary representation F_G, then for each natural number n we have

$$\frac{1}{|G|}\sum_{g\in G}\chi^{F_G}(g)^n, \quad \text{as well as} \quad \frac{1}{|G|}\sum_{g\in G}\chi^{F_G}(g^n)$$

are rational integral.

Proof: Since

$$\frac{1}{|G|}\sum_{g\in G}\chi^{F_G}(g)^n = (\overset{n}{\otimes} F_G, IG),$$

the first part is trivial.

For the second part we use 3.27 (i) which implies

$$\frac{1}{|G|}\sum_{g\in G}\chi^{F_G}(g^n) = \sum_{\substack{\alpha\vdash n\\ n_\alpha>0}}\zeta^\alpha((1\ldots n))\frac{1}{|G|}\sum_{g\in G}\chi^{F_G\boxdot[\alpha]}(g).$$

We know, that

$$\frac{1}{|G|}\sum_{g\in G}\chi^{F_G\boxdot[\alpha]}(g) = (F_G\boxdot[\alpha], IG) \in \mathbb{Z},$$

as well as that ζ^α has all its values in $\mathbb{Z}$ ($\mathbb{Q}$ is a splitting field, cf. I).

Having proved this, van Zanten and de Vries consider the result of applying the mapping

3.29 $\quad \sigma : G \to G : g \mapsto g^n$, $n \in \mathbb{N}$ fixed and prime to $|G|$,

which induces a permutation of the ordinary irreducible

characters of G, so that together with F_G there is an ordinary irreducible representation F_G^σ of G with character

3.30 $$\forall g \in G\ (\chi^{F_G^\sigma}(g) := \chi^{F_G}(g^n)).$$

Van Zanten and de Vries proved that

3.31 <u>If G is a finite group with an ordinary irreducible representation F_G and if n, σ and F_G^σ are as is described in 3.29, 3.30, then $\overset{n}{\otimes} F_G$ contains F_G^σ as an irreducible constituent.</u>

<u>Proof:</u> Otherwise we have (use 3.27 (ii)):

$$0 = \frac{1}{|G|}\sum_{g\in G}\chi^{F_G}(g)^n\overline{\chi^{F_G}(g^n)}$$

$$= \frac{1}{|G|}\sum_{\substack{\alpha\vdash n\\ n_\alpha>0}} f^\alpha \sum_{g\in G}\chi^{F_G\boxdot[\alpha]}(g)\overline{\chi^{F_G}(g^n)},$$

what implies, that for each partition α of n with $n_\alpha>0$:

$$\frac{1}{|G|}\sum_{g\in G}\chi^{F_G\boxdot[\alpha]}(g)\overline{\chi^{F_G}(g^n)} = 0.$$

Thus also

$$\frac{1}{|G|}\sum_{g\in G}\overline{\chi^{F_G}(g^n)}\sum_{\substack{\alpha\vdash n\\ n_\alpha>0}}\zeta^\alpha((1\ldots n))\chi^{F_G\boxdot[\alpha]}(g) = 0$$

so that an application of 3.27 (i) gives us

$$\frac{1}{|G|}\sum |\chi^{F_G}(g^n)|^2 = 0,$$

which is impossible.

q.e.d.

For further results concerning the number of solutions of the equation 3.25 as well as the integers

$$\frac{1}{|G|}\sum_{g \in G} \chi^{F_G}(g)^n \quad \text{and} \quad \frac{1}{|G|}\sum_{g \in G} \chi^{F_G}(g^n),$$

the reader may consult the paper of van Zanten and de Vries.

We have got some nice results about characters of symmetrized inner products, but there are still some fundamental questions to be answered, for example we would like to know at least some special cases, where $F_G \boxdot D^i$ is irreducible and we would prefer a more direct answer than 3.16 to the question of the existence of $F_G \boxdot D^i$.

These problems can be attacked by looking closer at the definition of $F_G \boxdot D^i$.

The definition was, that the H-invariant subspace of $\overset{n}{\otimes} V$, which affords n_i-times the representation D^i of H ($n_i := (\check{F}_G, D^i)$) affords f^i-times (f^i := dimension of D^i) a certain representation of G, and this representation was denoted by $F_G \boxdot D^i$. In other words: <u>The direct decomposition of $\overset{n}{\otimes} V$ into its homogeneous components with respect to H is also a direct decomposition of $\overset{n}{\otimes} V$ with respect to G,</u>

<u>and the homogeneous component of type D^i of $\overset{n}{\otimes} V$ with respect to H yields a left G-module which splits into f^i pairwise equivalent left G-modules. The representation afforded by each one of these direct summands is denoted by $F_G \boxdot D^i$.</u>

It can be shown (cf. Boerner [1]) that $F_G \boxdot D^i$ is afforded by $e^i(\overset{n}{\otimes}V)$, if e^i denotes a primitive idempotent of KH such that KHe^i affords D^i. Let us furthermore denote the centrally primitive idempotent which generates the homogeneous component of D^i in KH by e_i, i.e. we put

3.32 $$e_i := \frac{f^i}{|H|} \sum_{\pi \in H} \chi^{D^i}(\pi^{-1})\pi.$$

Hence the homogeneous decomposition of $\overset{n}{\otimes} V$ with respect to H is

3.33 $$\overset{n}{\otimes} V = \bigoplus_i (e_i \overset{n}{\otimes} V).$$

The sum can be taken over a complete system of pairwise inequivalent and irreducible K-representations of H, since in the case $n_i := (\check{F}_G, D^i) = 0$ we have $e_i \overset{n}{\otimes} V = \{0\}$. Thus

3.34 <u>$F_G \boxdot D^i$ exists if and only if $e_i \overset{n}{\otimes} V \neq \{0\}$.</u>

Let us consider the case $K := \mathbb{C}$, $H := S_n$.

3.16 has shown, that $F_G \boxdot [n]$ exists. What about $F_G \boxdot [1^n]$?

3.35 <u>If F_G is an ordinary representation of dimension f^{F_G}</u>

<u>of a group G, then $F_G \boxdot [1^n]$ exists if and only if $n \leq f^{F_G}$. The dimension of $F_G \boxdot [1^n]$ is $\binom{f^{F_G}}{n}$ if it exists.</u>

<u>Proof:</u> The centrally primitive idempotent corresponding to $[1^n]$ is

3.36 $$e_{(1^n)} = \frac{1}{n!} \sum_{\pi \in S_n} \epsilon_\pi \pi .$$

Hence $F_G \boxdot [1^n]$ exists if and only if

3.37 $$\left(\sum_{\pi \in S_n} \epsilon_\pi \pi\right) \overset{n}{\otimes} V \neq \{0\} .$$

If $\{e_1, \ldots, e_{f^{F_G}}\}$ is a basis of V, then

$$\{e_{i_1} \otimes \ldots \otimes e_{i_n} \mid 1 \leq i_\nu \leq f^{F_G}\}$$

is a basis of $\overset{n}{\otimes} V$, so that $F_G \boxdot [1^n]$ does not exist if and only if for each such basis vector $\underset{\nu}{\otimes} e_{i_\nu}$ we have

3.38 $$\left(\sum_{\pi \in S_n} \epsilon_\pi \pi\right) \underset{\nu}{\otimes} e_{i_\nu} = \sum_{\pi \in S_n} \epsilon_\pi \underset{\nu}{\otimes} e_{i_{\pi^{-1}(\nu)}} = 0.$$

(i) If $f^{F_G} < n$, then in each such basis vector $\underset{\nu}{\otimes} e_{i_\nu}$ at least two factors e_{i_ν} are equal, say e_{i_k} and e_{i_l}. Thus the transposition (kl) satisfies

$$\forall \pi \in S_n \left(\pi \underset{\nu}{\otimes} e_{i_\nu} = \pi(kl) \underset{\nu}{\otimes} e_{i_\nu}\right). \qquad (1)$$

Since

$$\sum_{\pi \in S_n} \epsilon_\pi \pi = -\sum_{\pi \in S_n} \epsilon_\pi \pi(kl),$$

we see, that (1) implies $(\Sigma\, \epsilon_\pi \pi) \overset{n}{\otimes} V = \{0\}$.

(ii) If on the other hand $f^{F_G} \geq n$, then there are $f^{F_G}(f^{F_G}-1)\ldots(f^{F_G}-n+1)$ basis vectors $\otimes e_{i_\nu}$ with pairwise different factors e_{i_ν}, and all the other basis vectors are mapped onto the zero vector by left multiplication (see (i)).

The basis vector $\otimes e_{i_\nu}$ with pairwise different factors e_{i_ν} is mapped onto

$$\sum_{\pi \in S_n} \epsilon_\pi \underset{\nu}{\otimes} e_{i_{\pi^{-1}(\nu)}} \in \overset{n}{\otimes} V$$

as are all the other basis vectors with the same <u>set</u> $\{e_{i_\nu} \mid 1 \leq \nu \leq n\}$ of factors up to a sign ± 1. Hence the image space has dimension $\binom{f^{F_G}}{n}$:

3.39 $$\Big(\sum_{\pi \in S_n} \epsilon_\pi \pi \overset{n}{\otimes} V : \mathbb{C}\Big) = \binom{f^{F_G}}{n} > 0, \quad \text{if} \quad f^{F_G} \geq n.$$

q.e.d.

<u>3.40</u> <u>If F_G is an ordinary representation of dimension f^{F_G} of a group G and $\alpha = (\alpha_1,\ldots,\alpha_h) \vdash n$, then $F_G \boxdot [\alpha]$ exists if and only if $h \leq f^{F_G}$.</u>

<u>Proof</u>: This theorem contains 3.35 and it is surprising, that it is proved mainly by an application of 3.35.

To show this we use the fact that the centrally primitive idempotent e_α corresponding to $[\alpha]$ can be expressed as a sum of orthogonal primitive idempotents, each of which generates a minimal left ideal in CS_n which affords $[\alpha]$. Since all these minimal left ideals which afford $[\alpha]$ are isomorphic to $CS_n e_1^\alpha$ (recall from part I that $e_1^\alpha = \mathcal{H}_1^\alpha \mathcal{V}_1^\alpha$), we need only show that the following is valid:

3.41 $$\mathcal{H}_1^\alpha \mathcal{V}_1^\alpha \overset{n}{\otimes} V \neq \{0\} \Leftrightarrow h \leq f^{F_G}.$$

To prove this, we use the fact that if $V_{\alpha'} = \times_j S_{\alpha'_j}$ then

$$\mathcal{V}_1^\alpha = \prod_j \left(\sum_{\pi \in S_{\alpha'_j}} \epsilon_\pi \pi \right),$$

and that the factors on the right hand side of this equation commute, so that

$$\mathcal{V}_1^\alpha = \prod_{j \geq 2} \left(\sum_{\pi \in S_{\alpha'_j}} \epsilon_\pi \pi \right) \left(\sum_{\rho \in S_{\alpha'_1}} \epsilon_\rho \rho \right).$$

3.35 yields, that

3.42 $$\left(\sum_{\rho \in S_{\alpha'_1}} \epsilon_\rho \rho \right) \overset{n}{\otimes} V \neq \{0\} \Leftrightarrow h \leq f^{F_G}.$$

And the argument in the proof of 3.35 shows clearly that this is equivalent to 3.41.

q.e.d.

What can be said about $F_G \boxdot D^i$, if D^i is an ordinary irreducible representation of a subgroup H of S_n ?

$F_G \boxdot D^i$ exists if and only if $e_i \overset{n}{\otimes} V \neq \{0\}$, i.e. if and only if D^i occurs under the irreducible constituents of $\check{F}_G$. $\check{F}_G$ is just the restriction of a representation of S_n on $\overset{n}{\otimes} V$, so that D^i occurs if and only if there is an occuring $[\alpha]$ which satisfies

3.43 $$([\alpha] \downarrow H, D^i) > 0.$$

Hence we have

3.44 $$(\check{F}_G, D^i) = \sum_{\substack{\alpha \vdash n \\ n_\alpha > 0}} ([\alpha] \downarrow H, D^i) n_\alpha.$$

This reduces the problem to a question about the relationship of the representation theories of S_n and H:

<u>3.45</u> <u>If F_G is an ordinary representation of a group G and if H is a subgroup of S_n with an ordinary irreducible representation D^i, then $F_G \boxdot D^i$ exists if and only if there is a partition $\alpha = (\alpha_1, \ldots, \alpha_h)$ of n satisfying $h \leq f^{F_G}$ and $([\alpha] \downarrow H, D^i) > 0$. If this happens, the dimension of $F_G \boxdot D^i$ is</u>

$$\sum_{\substack{\alpha \vdash n \\ h \leq f^{F_G}}} ([\alpha] \downarrow H, D^i) n_\alpha.$$

In the case when D^i, D^k are irreducible representations of H over an algebraically closed field K with $\mathrm{char}K \nmid |H|$ and so that $F_G \boxdot D^i$ as well as $F_G \boxdot D^k$ exist, for a given K-representation F_G of a group G, then it is of course reasonable to define.

$$F_G \boxdot (D^i + D^k) := F_G \boxdot D^i + F_G \boxdot D^k.$$

This may be generalized as follows:

<u>3.46 Def.:</u> If G is a group, H a subgroup of S_n, K an algebraically closed field with $\mathrm{char}K \nmid |H|$ and F_G, F_H are two K-representations (maybe reducible) of G, H, respectively, so that for at least one irreducible constituent D^i of F_H the symmetrized inner product exists, then we put

$$F_G \boxdot F_H := \sum_i (F_H, D^i)(F_G \boxdot D^i)$$

summing over all i for which $F_G \boxdot D^i$ exists (i the indices of a complete set of pairwise inequivalent irreducible constituents D^i of F_H).

This allows us to state the following theorem (Clausen [1]):

<u>3.47</u> <u>If F_G is an ordinary representation of dimension f^{F_G} of a group G, and D^i an ordinary irreducible representation of $H \leq S_n$, then, if $F_G \boxdot D^i$ exists, we have</u>

$$F_G \boxdot D^i = \sum_{\substack{\alpha \vdash n \\ h \leq f^{F_G}}} ([\alpha] \downarrow H, D^i) F_G \boxdot [\alpha].$$

<u>Proof:</u> $e^i \overset{n}{\otimes} V = e^i (\underset{\alpha \vdash n}{\oplus} (e_\alpha \overset{n}{\otimes} V)) = \underset{\alpha \vdash n}{\oplus} e_i e_\alpha (\overset{n}{\otimes} V)$

$$= \underset{\alpha \vdash n}{\oplus} ([\alpha] \downarrow H, D^i)(e_1^\alpha \overset{n}{\otimes} V).$$

q.e.d.

Hence the decomposition

3.48 $$\overset{n}{\otimes} F_G = \sum_{\substack{\alpha \vdash n \\ h \leq f^{F_G}}} f^\alpha (F_G \boxdot [\alpha])$$

into the symmetrized products with ordinary irreducible representations of S_n is at least as fine as is the decomposition into symmetrized products with ordinary irreducible representations of any $H \leq S_n$.

To the remaining question, which of $F_G \boxdot D^i$ are irreducible, I only mention that in the case $G := GL(m,\mathbb{C})$, $H := S_n$, $K := \mathbb{C}$ and

$$F_G := id_{GL(m,\mathbb{C})} : GL(m,\mathbb{C}) \to GL(m,\mathbb{C}) : g \mapsto g,$$

each existing $F_G \boxdot [\alpha]$ is irreducible, i.e. that we have

3.49 $\alpha = (\alpha_1, \ldots, \alpha_h) \vdash n$ and $h \leq m \Rightarrow \langle\alpha\rangle := id_{GL(m,\mathbb{C})} \boxdot [\alpha]$ is irreducible.

This is one of the crucial facts concerning the connection between the representation theories of symmetric and general linear groups. The reason for this remarkable result is that to this representation correspond two subalgebras of $\mathrm{End}_{\mathbb{C}}(\overset{n}{\otimes}\,\mathbb{C}^m)$, one of them a homomorphic image of $\mathbb{C}S_n$, the other a homomorphic image of $\mathbb{C}GL(m,\mathbb{C})$ and so that the latter is the enveloping algebra of the centralizer of the former (cf. Boerner [2], Clausen [1], Curtis/Reiner [1], Dieudonné/Carrell [1], Weyl [4]). This fact and various consequences are well known (cf. the quoted literature) and basic for the methods which especially D.E. Littlewood uses in his development of the representation theory of general linear and symmetric groups. They will be described in detail in another part of these lecture notes.

The main tool of Littlewood is the consequent use of certain polynomials which are closely related to the characters of the representation $\langle\alpha\rangle$.

3.22 yields for the character $\zeta^{\langle\alpha\rangle}$ of the representation $\langle\alpha\rangle$ of $GL(m,\mathbb{C})$:

<u>3.50</u> <u>If $m,n \in \mathbb{N}$, $\alpha := (\alpha_1,\ldots,\alpha_h) \vdash n$, $h \leq m$, then</u>

$$\forall\, g \in GL(m,\mathbb{C})\Big(\zeta^{\langle\alpha\rangle}(g) = \frac{1}{n!}\sum_{\pi\in S_n} \zeta^{\alpha}(\pi)\prod_{k=1}^{n} \mathrm{tr}(g^k)^{a_k(\pi)}\Big).$$

The trace $\mathrm{tr}(g^k)$ is the value of the polynomial

$$\sigma_{k,m} := \sum_{i=1}^{m} x_i^k \in \mathbb{C}[x_1,\ldots,x_m]$$

at $(\epsilon_1,\ldots,\epsilon_m)$, where ϵ_i are the characteristic roots of the matrix g.

Hence if we put

3.51
$$\{\alpha\} := \frac{1}{n!} \sum_{\pi\in S_n} \zeta^{\alpha}(\pi) \prod_{k=1}^{n} \sigma_{k,m}^{a_k(\pi)}$$

$$= \sum_{a\vdash n} \zeta_a^{\alpha} \prod_{k=1}^{n} \frac{1}{k^{a_k}\cdot a_k!} \sigma_{k,m}^{a_k},$$

where the sum is taken over all types $a = (a_1,\ldots,a_n)$ of n (for short: $a\vdash n$) and where ζ_a^{α} denotes the value of the ordinary irreducible character ζ^{α} (of $[\alpha]$) of S_n on the conjugacy class of elements of type a. These polynomials are called Schur-functions (for short: S-functions). We shall meet these polynomials again in a seemingly quite different region of mathematics, namely the theory of enumeration under group action which will be discussed in the following sections. Other expressions of $\{\alpha\}$ can be used by comparing them with 3.51 in order to obtain polynomials the coefficients of which turn out to be character values. It can be shown for example, that the following holds (cf. Boerner [2]):

3.52
$$\{\alpha\} = \frac{\det(x_i^{\alpha_j+m-j})}{\det(x_i^{m-j})}.$$

Comparing this with 3.51 we obtain a famous formula:

3.53 ("Frobenius' formula")

$$\frac{\det(x_i^{\alpha_j+m-j})}{\det(x_i^{m-j})} = \sum_{a \vdash n} \zeta_a^{\alpha} \prod_{k=1}^{n} \frac{1}{k^{a_k} \cdot a_k!} \sigma_{k,m}^{a_k} .$$

If we take e.g. m := 3 and α := (2,1) we obtain the polynomial

$$x_1^2x_2+x_2^2x_1+x_2^2x_3+x_1x_3^2+x_1^2x_3+x_2x_3^2+2x_1x_2x_3 = \frac{1}{3}(\sigma_1^3 - \sigma_3) \qquad (1)$$

The right hand side of 3.53 is

$$\frac{1}{6}\zeta_{(3,0,0)}^{(2,1)}\sigma_1^3 + \frac{1}{2}\zeta_{(1,0,0)}^{(2,1)}\sigma_1\sigma_2 + \frac{1}{3}\zeta_{(0,0,1)}^{(2,1)}\sigma_3 . \qquad (2)$$

Since the power sums $\sigma_{1,m},\ldots,\sigma_{n,m}$ are algebraically independent, if $n \leq m$, a comparison of (1) and (2) yields:

$$\zeta_{(3,0,0)}^{(2,1)} = 2, \quad \zeta_{(1,1,0)}^{(2,1)} = 0, \quad \zeta_{(0,0,1)}^{(2,1)} = 1.$$

This shows, how S-funktions can be used in order to evaluate characters of symmetric groups. This is useful and important, but the main thing is, that the mapping

$$[\alpha] \to \{\alpha\}$$

can be extended to an isomorphism from a ring which contains all the ordinary representations of all finite symmetric groups onto a ring of symmetric functions. Hence in order to derive results on ordinary representations of symmetric groups one is free in considering the representations in question or the corresponding polynomials. D.Knutson gave a modern treatment of this and called this result the "Fundamental Theorem of the Representation Theory of the Symmetric Group" (Knutson [1], p. 135).

But let us continue this section with a fascinating and very surprising application of 3.47. In fact we would like to show using

3.47 that the ring of n×n matrices over $\mathbb{C}$ satisfies the standard polynomial identity of degree 2n. B. Kostant was the first to show that this can be done by an application of symmetrization (cf. Kostant [1]) and I think that in particular this example shows that symmetrization is in fact very useful in order to discover deep symmetries in various mathematical structures.

We apply 3.47 to the case

$$G := GL(n,\mathbb{C}),\ F_G := id_{GL(n,\mathbb{C})},\ H := A_{2n+1},\ D^i := [n+1,1^n]$$

(recall that $[n+1,1^n]\downarrow A_{2n+1} = [n+1,1^n]^+ + [n+1,1^n]^-$, cf. I 4.54).

We state that the following holds (cf. Amitsur/Levitzki [1]):

<u>3.54</u> <u>If $A_1,\ldots,A_{2n}$ are n×n-matrices over $\mathbb{C}$, then they satisfy the following "standard polynomial identity":</u>

$$[A_1,\ldots,A_{2n}] := \sum_{\pi\in S_{2n}} \epsilon_\pi A_{\pi(1)} \cdots A_{\pi(2n)} = 0.$$

<u>Proof</u>:

(i) We first prove that the mapping

$$\sum_{\pi\in S_{2n+1}} \epsilon_\pi (e;\pi(1\ldots 2n+1)\pi^{-1}): \overset{2n+1}{\otimes} \mathbb{C}^n \to \overset{2n+1}{\otimes} ,$$

which maps $v_1\otimes\ldots\otimes v_{2n+1}$ onto

$$\sum_{\pi\in S_{2n+1}} \epsilon_\pi (\ldots\otimes v_{\pi(2n+1\ldots 1)\pi^{-1}(i)} \otimes\ldots)$$

is the zero mapping.

In order to prove this, we notice first that $id_{GL(n,\mathbb{C})} \boxdot [n+1,1^n]$ does not exist (use 3.40). Hence 3.47 yields that both

$id_{GL(n,\mathbb{C})} \boxdot [n+1,1^n]^{\pm}$ do not exist. Expressed in terms of the corresponding centrally primitive idempotents $e_{[n+1,1^n]^{\pm}}$ this reads as follows:

$$e_{[n+1,1^n]^{\pm}} \overset{2n+1}{\otimes} \mathbb{C}^n = \{0\},$$

so that we also have

$$(e_{[n+1,1^n]^+} - e_{[n+1,1^n]^-}) \overset{2n+1}{\otimes} \mathbb{C}^n = \{0\}$$

Applying Frobenius' theorem on the characters of alternating groups (I. 4.55) to this situation we see that this difference of idempotents is a multiple of

$$\sum_{\rho\in(2n+1)^+} \rho - \sum_{\tau\in(2n+1)^-} \tau = \lambda \Big(\sum_{\pi\in S_{2n+1}} \epsilon_\pi \pi(1\ldots 2n+1)^{-1}\pi^{-1} \Big)$$

($(2n+1)^{\pm}$ the two A_{2n+1}-classes into which the class of $(1\ldots 2n+1)$ splits). This yields our first statement.

(ii) We now prove that $tr[A_1,\ldots,A_{2n+1}] = 0$ for arbitrary nxn-matrices over $\mathbb{C}$.

$$tr[A_1,\ldots,A_{2n+1}] = tr \sum_{\pi\in S_{2n+1}} \epsilon_\pi A_{\pi(1)} \cdots A_{\pi(2n+1)}$$

$$= \sum \epsilon_\pi \, tr\, A_{\pi(1)} \cdots A_{\pi(2n+1)}$$

$$= \sum \epsilon_\pi \, tr\, \widehat{\#id_{End(V)}}^{2n+1}(A_1,\ldots,A_{2n+1};\pi(1,\ldots,2n+1)^{-1}\pi^{-1})$$

(cf. proof of 2.5). Hence

$$tr[A_1,\ldots,A_{2n+1}] = tr\widehat{\#id_{End(V)}}^{2n+1}((A_1,\ldots,A_{2n+1};1)\circ\sum_\pi \epsilon_\pi(e;\pi(1..2n+1)^{-1}\pi^{-1}))$$

and this is 0 since (i).

(iii) We are now in the position to prove the statement. In order to do this we first remark that

$$S_{2n+1} = S_{2n} \cdot \langle (1 \ldots 2n+1) \rangle$$

(S_{2n} may be considered as stabilizer of the symbol $2n+1$ in S_{2n+1}). This together with (ii) gives us

$$0 = tr[A_1, \ldots A_{2n+1}] = \sum_{\pi \in S_{2n+1}} \epsilon_\pi \, tr \, A_{\pi(1)} \cdots A_{\pi(2n+1)}$$

$$= \sum_{\sigma \in S_{2n}} \epsilon_\sigma \sum_{\tau \in \langle (1 \ldots 2n+1) \rangle} \epsilon_\tau \, tr \, A_{\sigma\tau(1)} \cdots A_{\sigma\tau(2n+1)}$$

$$= (2n+1) \, tr \, A_{2n+1} \cdot \sum_{\sigma \in S_{2n}} \epsilon_\sigma A_{\sigma(1)} \cdots A_{\sigma(2n)}$$

$$= (2n+1) \, tr \, A_{2n+1} \cdot [A_1, \ldots, A_{2n}].$$

Since this holds for arbitrary $n \times n$-matrices A_{2n+1}, the statement must be satisfied.

q.e.d.

After having considered this interesting application, we are left with the question how symmetrized inner products split. We would like to discuss this question for the case when both factors are ordinary irreducible representations of symmetric groups, i.e. we ask for the splitting of $[\alpha] \boxdot [\beta]$ into its irreducible consituents.

This problem has not been solved completely but there is a formula which reduces this problem to the splitting of symmetrized inner products of form $[\alpha] \boxdot [r]$, $r \leq n$, if $\beta \vdash n$. This formula is quite analogous to formula I. 6.20, where the splitting of outer symmetrized products $[\alpha] \odot [\beta]$ was reduced to the problem to split products of form $[\alpha] \odot [r]$.

We first state that the following holds (cf. Robinson [5], 3.62):

3.55 $([\alpha] \boxdot [n_1]) \otimes ([\alpha] \boxdot [n_2]) = [\alpha] \boxdot ([n_1][n_2])$.

Proof: If $\alpha \vdash m$, $\pi \in S_m$, then the character of the left hand side is

$$\chi^{[\alpha] \boxdot [n_1]}(\pi)\, \chi^{[\alpha] \boxdot [n_2]}(\pi) =$$

$$\frac{1}{n_1!} \sum_{\rho \in S_{n_1}} \prod_{k=1}^{n_1} \zeta^{\alpha}(\pi^k)^{a_k(\rho)} \cdot \frac{1}{n_2!} \sum_{\sigma \in S_{n_2}} \prod_{l=1}^{n_2} \zeta^{\alpha}(\pi^{l})^{a_l(\sigma)} =$$

$$\frac{1}{n_1!n_2!} \sum_{\substack{\tau = \rho\sigma \\ \in S_{n_1} \times S_{n_2}}} \prod_{k=1}^{n_1+n_2} \zeta^{\alpha}(\pi^k)^{a_k(\tau)} . \qquad (1)$$

The character of the right hand side is

$$\chi^{[\alpha] \boxdot ([n_1][n_2])}(\pi) = \frac{1}{(n_1+n_2)!} \sum_{\psi \in S_{n_1+n_2}} \chi^{[n_1][n_2]}(\psi) \prod_{k=1}^{n_1+n_2} \zeta^{\alpha}(\pi^k)^{a_k(\psi)}$$

$$= \frac{1}{(n_1+n_2)!} \sum_{\psi \in S_{n_1+n_2}} \frac{(n_1+n_2)!\ |C^{S_{n_1+n_2}}(\psi) \cap S_{n_1} \times S_{n_2}|}{n_1!n_2! \cdot |C^{S_{n_1+n_2}}(\psi)|} \prod_{k=1}^{n_1+n_2} \zeta^{\alpha}(\pi^k)^{a_k(\psi)}$$

This is obviously equal to (1). q.e.d.

This formula yields at once a determinantal expression for $[\alpha] \boxdot [\beta]$. If instead of the multiplication $\cdot$ we use inner tensor product multiplication (we indicate this by writing $|\ -\ |^{\otimes}$)

we obtain (Robinson [5], 3.63):

3.56 $$[\alpha] \boxdot [\beta] = |[\alpha] \boxdot [\beta_i+j-i]|^{\otimes} .$$

Hence for example

$$[\alpha] \boxdot [2,1] = \begin{vmatrix} [\alpha] \boxdot [2] & [\alpha] \boxdot [3] \\ 1 & [\alpha] \boxdot [1] \end{vmatrix}^{\otimes}$$

$$= ([\alpha] \boxdot [2]) \otimes ([\alpha] \boxdot [1]) - [\alpha] \boxdot [3]$$

$$= ([\alpha] \boxdot [2]) \otimes [\alpha] - [\alpha] \boxdot [3].$$

In order to prove 3.56 one needs in fact a result which is more general than 3.55. One needs

3.57 $$([\alpha] \boxdot [\beta]) \otimes ([\alpha] \boxdot [\gamma]) = [\alpha] \boxdot ([\beta][\gamma]).$$

This fact can be proved in the same way as 3.55. The corresponding character formulae are longer winded hence I left it out. If we agree upon 3.57, then 3.56 is obtained by applying I 4.41 which yields

$$[\beta] = |[\beta_i + j - i]|.$$

There are various other results on the decomposition of $[\alpha] \boxdot [\beta]$.

If F denotes an ordinary irreducible representation of a finite group G, then we may put the question, whether $F \boxdot [2]$ and $F \boxdot [1^2]$ contain the identity representation IG of G. It is interesting to notice that an answer to this question characterizes the kind of F (one calls F to be a representation of the first kind if and only if F is equivalent to a real representation, if this is not the case then F is called a representation of the second or third kind, respectively, if and only if its character ζ^F has real values only or not, respectively).

For a well known result says that the following holds. If we put

3.58 $$c_F := \frac{1}{|G|} \sum_{g \in G} \zeta^F(g^2),$$

then we have (cf. e.g. Feit [1], (3.5)):

3.59 F is of the first, second, third kind, respectively, if and only if $c_F = 1, -1, 0$, respectively.

The complete proof of 3.59 does not fit in very nicely in this context. But we can show at least that c_F is in fact an invariant and how it can be expressed in terms of symmetrized products. If F^c denotes the representation contragredient to F, i.e. if we put

3.60 $$\forall\, g \in G\ (F^c(g) := {}^tF(g^{-1}))$$

(${}^tF(g^{-1})$) the transpose of $F(g^{-1})$), then we have for its character:

3.61 $$\forall\, g \in G\ (\zeta^{F^c}(g) = \overline{\zeta^F(g)}\ ).$$

We notice first that

3.62 $(F,F^c) = (F \otimes F,\ IG)$,

so that we obtain in terms of symmetrized products:

3.63 $(F,F^c) = (F \boxdot [2],\ IG) + (F \boxdot [1^2],\ IG) \in \{0,1\}$.

This implies (use 3.24 (ii) and (iii)):

3.64 $c_F = (F \boxdot [2],\ IG) - (F \boxdot [1^2],\ IG) \in \{0,1,-1\}$.

3.61 shows that F is of the third kind if and only if $(F,F^c) = 0$ so that 3.63 yields

3.65 F is of the third kind if and only if

$0 = (F \boxdot [2],\ IG) = (F \boxdot [1^2],\ IG)\ \left(\Leftrightarrow \frac{1}{|G|} \sum_g \zeta^F(g^2) = 0\right)$.

If F is of the first kind, then it is well known that we may assume that the representing matrices are orthogonal, so that $\mathbb{F}(g) = \mathbb{F}^c(g)$. This yields (we put $\mathbb{F}(g) = (f_{ik}(g))$):

$$\begin{aligned} c_F &= \frac{1}{|G|} \sum_{g \in G} \zeta^F(g^2) = \frac{1}{|G|} \sum_{g \in G} \operatorname{tr}(\mathbb{F}(g)\ \mathbb{F}(g)) \\ &= \frac{1}{|G|} \sum_{g \in G} \sum_{i,k} f_{ik}(g)\ f_{ik}\ (g^{-1}) \\ &= \frac{1}{|G|} \sum_{i,k} \sum_{g \in G} f_{ik}(g)\ f_{ik}\ (g^{-1}) \\ &= 1. \end{aligned}$$

The last equation follows from the orthogonality relations (Feit [1], (1.9)). It is not as easy as that to show that second kind implies $c_F = -1$. A corollary of 3.58 - 3.65 is:

3.66 Let F denote an ordinary irreducible representation of a finite group. Then

(i) F is equivalent to a real representation if and only if $(F \boxdot [2], IG) = 1$, and $(F \boxdot [1^2], IG) = 0$.

(ii) F is not equivalent to a real representation but has a real-valued character if and only if $(F \boxdot [2], IG) = 0$, and $(F \boxdot [1^2], IG) = 1$.

(iii) The character of F is not real-valued if and only if $(F \boxdot [2], IG) = (F \boxdot [1^2], IG) = 0$.

This is a characterization of the kind of F in terms of multiplicities of I_G in the inner symmetrized products $F \boxdot [2]$ and $F \boxdot [1^2]$.

Since $\mathbb{Q}$ is a splitting field for S_n, we obtain e.g.

3.67 $\forall\, \alpha \vdash n\ (([\alpha] \boxdot [2], [n]) = 1 \wedge ([\alpha] \boxdot [1^2], [n]) = 0)$.

We have seen that a finite group G has $\mathbb{R}$ as splitting field if and only if for each of its ordinary irreducible representations F the following holds:

$$(F \boxdot [2], IG) \neq 0.$$

Hence it might be useful to ask for the multiplicity of the identity representation in various symmetrized inner products. Several physicists pointed to groups G with the property that for each of their ordinary irreducible representations F, IG is not contained in $F \boxdot [2,1]$. They called such groups <u>simple-phase groups</u> (van Zanten/de Vries [1], Butler [1], Butler/King [3]). Groups which are not simple-phase groups are called <u>non-simple-phase groups.</u> Correspondingly an ordinary irreducible representation F is called a <u>simple-phase representation</u> of G if and only if

$$(F \boxdot [2,1], IG) = 0,$$

and a <u>non-simple-phase representation</u> of G if and only if

$$(F \boxdot [2,1], IG) \neq 0.$$

An application of 3.24 (v) yields (for finite G):

<u>3.68</u> $m_F := (F \boxdot [2,1], IG) = \frac{1}{3|G|} \sum_{g \in G} (\zeta^F(g)^3 - \zeta^F(g^3))$

Let us consider the case $F := [\alpha]$, $\alpha \vdash n$.
An application of the example following 3.56 yields

$$m_{[\alpha]} = ([\alpha] \otimes ([\alpha] \boxdot [2]), [n]) - ([\alpha] \boxdot [3], [n]),$$

where we may substitute $[\alpha] \otimes [\alpha] - [\alpha] \boxdot [1^2]$ for $[\alpha] \boxdot [2]$ so that we obtain (regard 3.67):

3.69 $$m_{[\alpha]} = (\overset{3}{\otimes} [\alpha], [n]) - ([\alpha] \boxdot [3], [n]).$$

This shows, how the character table of S_n and the decomposition of $[\alpha] \boxdot [3]$ can be used in order to check whether $[\alpha]$ is simple-phase or not. It was shown by Derome (direct verification) that S_6 is a non-simple-phase group (cf. Derome [1]). Butler and King give many further examples (Butler/King [3]).

As far as I know the most complete tables with decompositions of symmetrized inner products $[\alpha] \boxdot [\beta]$ were given by Esper (Esper [2]), if $\alpha \vdash m$, $\beta \vdash n$, he gave such tables for all the partitions α and β with $2 \leq m \leq 10$ and $n \leq 5$ as well as for all α and β with $2 \leq m \leq 7$ and $n = 6$. Esper used the character formula 3.22, other people used various tricks like results concerning the decomposition of the corresponding symmetric functions into S-functions.

Concerning this, we should at least notice what follows like 3.57 for the corresponding S-functions.

<u>3.70</u> <u>The symmetric function corresponding to the representation $\langle[\beta][\gamma]\rangle = id_{GL(m,\mathbb{C})} \boxdot ([\beta][\gamma])$ is the product of the S-functions $\{\beta\}$ and $\{\gamma\}$.</u>

<u>Proof</u>: The symmetric function which corresponds to $\langle[\beta][\gamma]\rangle$ is

$$\frac{1}{(n_1+n_2)!} \sum_{\pi\in S_{n_1+n_2}} \frac{(n_1+n_2)!}{n_1!n_2!\,|C^{S_{n_1+n_2}}(\pi)|} \sum_{\rho\sigma\in C^{S_{n_1+n_2}}(\pi)\cap S_{n_1}\times S_{n_2}} \zeta^{\beta}(\rho)\,\zeta^{\gamma}(\sigma) \prod_{k=1}^{n_1+n_2} \sigma_{k,m}^{a_k(\pi)}$$

$$= \frac{1}{n_1!n_2!} \sum_{\rho\sigma\in S_{n_1}\dot{\times} S_{n_2}} \zeta^{\beta}(\rho)\,\zeta^{\gamma}(\sigma) \prod_{k=1}^{n_1+n_2} \sigma_{k,m}^{a_k(\rho)+a_k(\sigma)}$$

$$= \Big(\frac{1}{n_1!} \sum_{\rho\in S_{n_1}} \zeta^{\beta}(\rho) \prod_{k=1}^{n_1} \sigma_{k,m}^{a_k(\rho)}\Big)\Big(\frac{1}{n_2!} \sum_{\sigma\in S_{n_2}} \zeta^{\gamma}(\sigma) \prod_{k=1}^{n_2} \sigma_{k,m}^{a_k(\sigma)}\Big)$$

$$= \{\beta\}\{\gamma\}.$$

q.e.d.

This yields a result about S-functions which corresponds to I 4.41:

<u>3.71</u> $\underline{\{\alpha\} = |\{\alpha_i + j - i\}|.}$

(Where we put $\{0\} := 1$ and $\{m\} := 0$, if $m < 0$.)

I should not forget to mention that mostly 3.71 is proved first and then serves to obtain I 4.41 via the isomorphism described in the "Fundamental Theorem" which was mentioned in the remark following 3.53. (In fact 3.70 shows that the mapping is a multiplicative homomorphism so that together with the additivity (see the remark following 3.45) 3.70 and the bijectivity yield the "Fundamental Theorem".)

Chapter III

An application to combinatorics:

The theory of enumeration under group action

It is shown how characters of wreath products can be applied to the Redfield-Pólya enumeration theory. The evaluation of the characters of several permutation representations of $G \sim H$ allows to evaluate the number of orbits of these permutation groups as well as their cycle-indices. Furthermore it is shown how Schur-functions come in and how these can be used in enumeration theory.

4. Enumeration under group action

In mathematics as well as its applications we are quite often faced with the problem to count the number of equivalence classes of a given set under a given equivalence relation.

Illustrations of this situation are the so-called necklace-problems. A concrete example reads as follows. We are asked for the number of different necklaces with 5 beads in two colours. To make it more precise: We are asked for the number of ways in which the vertices of a pentagon can be coloured in this way. Two colourings of the pentagon are called different if and only if there is no element of the symmetry group D_5 of the pentagon which yields one of these two colourings from the other. The solution of this problem is 8, for if we denote by • and o the two colours, we obtain 8 different colourings as it is indicated in Fig. 1:

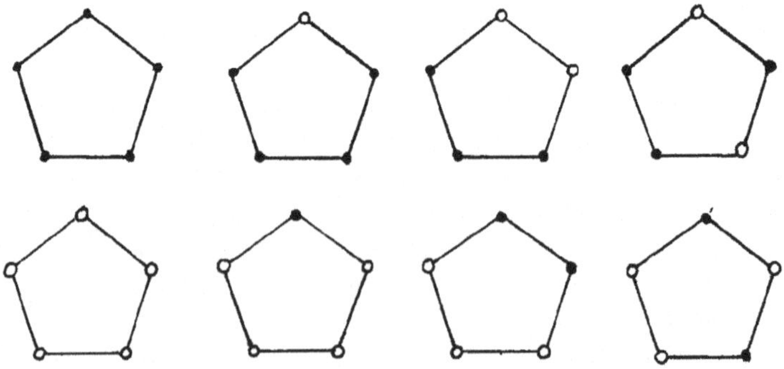

Fig. 1

Further problems of this kind and related problems are e.g.:
What is the number of colourings of the edges (vertices,faces) of the cube in 2 colours?

What is the number of labeled graphs of order n?
What is the number of chemical isomeres of a certain molecule, say of $C_n H_{2n+1} OH$?
What is the number of automata with n states, k input symbols, and m output symbols?
What is the number of elements of a given finite field K which are primitive with respect to a given subfield E of K?

The theory which has been developed in order to allow a unified treatment of this kind of problems is mainly the work of J.H. Redfield, G. Pólya and N.G. de Bruijn. It received an impetus especially when it was applied (in particular by F. Harary) to enumeration problems in graph theory during the last twenty years.

This theory originated in papers written by Cayley and several chemists, which contain results on the enumeration of organic molecules (cf. Cayley [1], [2] and the references in Pólya [1]).

J.H. Redfield was the first mathematician who wrote a paper on this subject where a certain polynomial is used, the so-called cycle-index of the permutation group considered, which is the main tool of this theory in its present form. This paper, published in 1927 and entitled "The theory of group reduced distributions",contains deep results and is not easy to read. This may be the reason for that it was overlooked for nearly thirty years. It was mentioned first by D.E. Littlewood in his book on group characters (Littlewood [2]), in the chapter where Littlewood discusses multiple transitivity of permutation groups.

Redfield discusses in his paper the use of symmetric functions for enumeration problems and my goal is now, to point to these results and to shed light upon the connection with the preceding results on representation theory of symmetric groups and wreath products. Hitherto several papers were published where Redfield's results and methods are discussed (Foulkes [1], [2], Harary/Palmer [3]). H.O. Foulkes in particular pointed to the close relationship with the representation theory of finite groups.

The second paper on that subject was published in 1937 by G. Pólya. Its title is "Kombinatorische Anzahlbestimmungen für Gruppen, Graphen und chemische Verbindungen" (Pólya [1]) and it contains both a systematic introduction and an excellent and extensive treatment of these enumeration techniques which were initiated by chemists and influenced by Cayley during the nineteenth century.

During the fifties of this century more and more of such enumeration problems were attacked. N.G. de Bruijn was the first to write a survey article on that subject (de Bruijn [1]), it was published in 1964 and is still a standard reference. N.G. de Bruijn obtained several generalizations of Pólya's results (cf. de Bruijn [4]).

This theory has meanwhile been applied by many authors to enumeration problems, especially in graph theory. A comprehensive treatment of these methods and results has recently been published by F. Harary and E. Palmer (Harary/Palmer [1]). Short and stimulating introductions to this theory were given by Harary as well as by de Bruijn (Harary/Beineke [1], de Bruijn [5]). The group theoretical aspects were pointed out in very interesting but hitherto unpublished notes by A. Rudvalis and E. Snapper (Rudvalis/Snapper [1]). Besides this, various textbooks on combinatorics contain chapters on enumeration under group action (cf. Berge [1], Liu [1]).

After this short sketch of the history of this theory let us return to our introductory example where we asked for the number of colourings of the regular pentagon by the colours ° and

Such a colouring of the pentagon may be regarded as a mapping from the set $\{1,\ldots,5\}$ of labeled vertices of the pentagon into the set $\{1,2\}$ of the two colours. Since we shall denote the set of all the mappings from a given domain D into the given range R by R^D as usual, the set

$$\{1,2\}^{\{1,\ldots,5\}} := \{\varphi \mid \varphi : \{1,\ldots,5\} \to \{1,2\}\}$$

may be regarded as the set of all the $2^5 = 32$ colourings of the pentagon by two colours.

It is clear that the solution 8 of the necklace-problem is the number of orbits of the permutation group induced by the symmetry group D_5 of the regular pentagon on the set $\{1,2\}^{\{1,\ldots,5\}}$.

The permutation group induced by D_5 on $\{1,2\}^{\{1,\ldots,5\}}$ is the image of the following permutation representation of D_5:

$$D_5 \ni \pi \mapsto (\varphi \mapsto \psi := \varphi \circ \pi^{-1})$$

Hence Burnside's lemma, which gives the number of orbits of a permutation group (in terms of its character) provides a solution of our problem.

Since all the enumeration theorems which will be mentioned in the sequel are in a sense generalizations of this lemma of Burnside, we shall now state this well-known result and we shall prove it by an argument which allows suitable generalizations later on:

4.1 ("Lemma of Burnside")

If P denotes a permutation group on a finite set X and if $a_1(p)$ denotes the number of points which remain fixed under $p \in P$, then the number of orbits of P on X is equal to

$$\frac{1}{|P|} \sum_{p \in P} a_1(p).$$

Proof: Instead of P we consider its natural representation ν on the $|X|$ - dimensional vector space $V := \mathbb{C}^{|X|}$ over $\mathbb{C}$ with basis $\{e_1,\ldots,e_{|X|}\}$:

$$\nu : P \to \mathrm{Aut}_{\mathbb{C}}(\mathbb{C}^{|X|}) : p \mapsto (e_i \mapsto e_{p(i)}).$$

If $\omega_1,\ldots,\omega_t$ are the orbits of P on X, then

$$V_i := \langle e_j \mid j \in \omega_i \rangle$$

is an invariant subspace.

The salient point is, that V_i contains exactly one subspace which affords the identity representation of P. (This holds since P acts transitively on $\{e_j | j \in \omega_i\}$ and the subspace is generated by $\sum_{j \in \omega_i} e_j$)

Hence the number of orbits of P on X is equal to the multiplicity of the identity representation in ν. But this multiplicity is equal to the dimension of the direct sum of subspaces of V which afford the identity representation. Furthermore the vector space V is projected onto this direct sum by left multiplication with the centrally primitive idempotent

$$\frac{1}{|P|}\sum_{p \in P} \nu(p).$$

The dimension of a projection is just its trace.

Hence we have obtained

4.2 no.of orbits of $P = (\nu P, IP) = \frac{1}{|P|}\sum_{p \in P} tr(\nu(p)) = \frac{1}{|P|}\sum_{p \in P} a_1(p)$.

q.e.d.

The symmetry group D_5 of the regular pentagon consists of 5 cyclic permutations:

1 permutation of type (5,0,0,0,0),
4 permutations of type (0,0,0,0,1),

and 5 reflections:

5 permutations of type (1,2,0,0,0).

It is clear that a colouring φ is fixed under $\pi \in D_5$ if and only if φ is constant on each cyclic factor. Hence the number of fixed points of the permutation induced by π on $\{1,2\}^{\{1,\ldots,5\}}$ is $2^{c(\pi)}$, where $c(\pi)$ again denotes the number of cyclic factors of π.

Thus 4.1 yields for the number of different necklaces:

$$\frac{1}{10}(1\cdot 2^5 + 4\cdot 2^1 + 5\cdot 2^3) = 8.$$

With this we have shown how enumeration problems of this special kind can be solved by an application of Burnside's lemma.

Let us restate this problem now in a more general form in order to describe generalizations. We are given two finite sets M and N, say. Without loss of generality we may assume that

$$M := \{1,\ldots,m\} \subseteq \mathbb{N} \supseteq N := \{1,\ldots,n\}.$$

We consider the set M^N of all the mappings φ from N to M:

$$M^N := \{\varphi \mid \varphi : N \to M\}.$$

With the aid of two permutation groups, G acting on M and H acting on N, a permutation group acting on M^N is defined and we are asked for a description of its orbits.

These orbits are called <u>patterns.</u>

There are three main types of patterns depending on the way how the permutation representation is defined:

4.3 Def.: (i) $G \leq S_m$, $H \leq S_n$, define the following permutation representation of $G \wr H$:

ρ: $G \wr H \to S_{M^N}$: $(f,\pi) \mapsto (\varphi \mapsto \psi,\ \psi(i) := f(i)\varphi(\pi^{-1}(i)))$.

$\rho[G \wr H]$, the image of this representation is called the exponentiation of G and H and denoted by $[G]^H$, i.e.

$$[G]^H := \rho[G \wr H].$$

(ii) The subgroup $\rho[\text{diag } G^{*}\cdot H']$ (recall that $\text{diag } G^{*}\cdot H' = \{(g,\ldots,g;\pi) \mid g \in G \wedge \pi \in H\}$) is called the power of G and H and denoted by G^H:

$$G^H := \rho[\text{diag } G^{*}\cdot H'] \leq \rho[G \wr H] = [G]^H$$

(iii) The subgroup $\rho[H'] = \rho[\{(e;\pi) \mid \pi \in H\}]$ is denoted by E^H:

$$E^H := \rho[H'] = \{1_{S_m}\}^H = [\{1_{S_m}\}]^H.$$

Notice that

4.4 $$E^H \leq G^H \leq [G]^H,$$

so that enumeration problems concerning G^H and E^H can be solved by restriction to suitable subgroups once we have solved the corresponding enumeration problem concerning $[G]^H$. Hence we would like to evaluate the number of orbits of $[G]^H$ first.

In order to do that we first characterize $[G]^H$ in terms of representation theory of wreath products (Kerber [8]):

<u>4.5</u> <u>If $G \leq S_m$, $H \leq S_n$, and ν denotes the natural representation of G, then $\widetilde{\overset{n}{\#}\,\nu}$ is a representation of $G \sim H$ with</u>

$$\nu[[G]^H] = \nu \circ \rho\,[G \sim H] = \widetilde{\overset{n}{\#}\,\nu}.$$

Proof: $\nu[G]^H$ acts on the vector space $\mathbb{C}^{m^n}$ which is isomorphic to $\overset{n}{\otimes}\,\mathbb{C}^m$, on which $\widetilde{\overset{n}{\#}\,\nu}$ acts. If $\{e_1, \ldots, e_m\}$ is a basis of $\mathbb{C}^m$, then

$$\{e_\varphi := e_{\varphi(1)} \otimes \ldots \otimes e_{\varphi(n)} \mid \varphi \in M^N\}$$

is a basis of $\overset{n}{\otimes}\,\mathbb{C}^m$, on which $(f;\pi) \in G \sim H$ acts as follows:

$$\begin{aligned}(f,\pi)\, e_\varphi &= f(1)\, e_{\varphi(\pi^{-1}(1))} \otimes \cdots \otimes f(n) e_{\varphi(\pi^{-1}(n))} \\ &= e_{f(1)\varphi(\pi^{-1}(1))} \otimes \cdots \otimes e_{f(n)\varphi(\pi^{-1}(n))} \\ &= e_\psi,\end{aligned}$$

if $\psi(i) := f(i)\, \varphi(\pi^{-1}(i))$, $1 \leq i \leq n$.

q.e.d.

This shows that the permutation character of $[G]^H$ is equal to the character of $\widetilde{\overset{n}{\#}\nu}$:

<u>4.6</u> $$\forall\, (f;\pi) \in G \sim H\ (a_1(\rho(f;\pi)) = \chi^{\widetilde{\overset{n}{\#}\,\nu}}(f;\pi)).$$

We need only apply 2.6 in order to get this character value explicitly:

<u>4.7</u> $$a_1(\rho(f;\pi)) = \chi^{\widetilde{\overset{n}{\#}\,\nu}}(f;\pi) = \prod_{i=1}^{c(\pi)} a_1(g_i(f;\pi)).$$

(This expression can be simplified a bit by using the type notation for the conjugacy classes.)

Burnside's lemma together with 4.7 yields now (Kerber [8]):

<u>4.8</u> <u>("Exponentiation group enumeration theorem, constant form")</u>

<u>The number of orbits of the exponentiation $[G]^H$ of $G \leq S_m$ and $H \leq S_n$ on M^N is equal to</u>

$$\frac{1}{|G|^n|H|} \sum_{(f;\pi)\in G \wr H} \prod_{i=1}^{c(\pi)} a_1\,(g_i(f;\pi)).$$

By restricting to the subgroup G^H we obtain the following theorem (de Bruijn [2], Harary/Palmer [2]):

<u>4.9</u> <u>(" Power group enumeration theorem, constant form"):</u>

<u>The number of orbits of the power G^H of $G \leq S_m$ and $H \leq S_n$ is equal to</u>

$$\frac{1}{|G||H|} \sum_{(g,\pi)\in G\times H} \prod_{i=1}^{n} \Big(\sum_{k|i} k\cdot a_k(g)\Big)^{a_i(\pi)}$$

<u>Proof:</u> If the j-th cyclic factor of π is of length r, then

$$g_j(f;\pi) = g^r$$

if the value of the constant mapping f is g.

In this case we have

$$a_1(g_j(f;\pi)) = a_1(g^r) = \sum_{s|r} s\cdot a_s(g).$$

This together with 4.8 yields the statement.

q.e.d.

Since

$$a_1(\rho(e;\pi)) = m^{c(\pi)}$$

we obtain for the subgroup E^H (Pólya[1]):

4.10 The number of orbits of E^H is equal to

$$\frac{1}{|H|}\sum_{\pi\in H} m^{c(\pi)}.$$

If we are interested in a numerical example we need evaluate the permutation character of $[G]^H$ so that the question arises for the decomposition of $\widetilde{\overset{n}{\#}\vee}$ into its irreducible constituents.

This decomposition depends heavily on G but the known character tables of wreath products are tables of wreath products of symmetric groups mostly (Sänger [1], Gretschel [1], Hilge [1]), so that we notice first that for the natural representations $\vee G$ of $G \leq S_m$ and $\vee S_m$ of S_m the following holds:

4.11 $$\underbrace{\widetilde{\overset{n}{\#}\vee G}}_{\text{of } G\wr H} = \underbrace{\widetilde{\overset{n}{\#}\vee S_m}}_{\text{of } S_m\wr S_n} \downarrow G\wr H$$

Hence we may ask for the decomposition of $\widetilde{\overset{n}{\#} \vee S_m}$, for which the following holds:

<u>4.12</u> $$\widetilde{\overset{n}{\#} \vee S_m} = \widetilde{\overset{n}{\#} ([m] + [m-1,1])}.$$

We apply 2.59 in order to derive that

<u>4.13</u> $$\widetilde{\overset{n}{\#}([m] + [m-1,1])} = \sum_{r=o}^{n} (m;r)\#(m-1,1;\, n-r) \uparrow S_m \wr S_n$$

Hence e.g.

$$\widetilde{\overset{3}{\#}([2]+[1^2])} = \widetilde{\overset{3}{\#}[1^2]} + \widetilde{([2]\#[1^2]\#[1^2]} \otimes ([1]\#[2])') \uparrow S_2 \wr S_3$$
$$+ \widetilde{([2]\#[2]\#[1^2]} \otimes ([2]\#[1])') \uparrow S_2 \wr S_3 + \widetilde{\overset{3}{\#}[2]}$$

The characters of these 4 ordinary irreducible representations of $S_2 \wr S_3$ can be found in the character table in Sänger [1]. An addition of their values shows that the resulting permutation character of $[S_2]^{S_3}$ has only 3 values $\neq 0$, these values and the orders and representatives of the corresponding classes in $S_2 \wr S_3$ are:

<u>4.14</u>

value	class order	representative of class
8	1	(e; 1)
4	6	(e; (12))
2	8	(e; (123))

An inspection of the conjugacy classes yields for the rectriction to $S_2^{S_3}$ the same character values, since (e; 1), (e; (12)) as well as (e; (123)) are contained in $S_2^{S_3}$. After a suitable change of class orders we obtain for the permutation character of $S_2^{S_3}$:

4.15

value	class order	representative
8	1	(e; 1)
4	3	(e; (12))
2	2	(e; (123))

The restriction to the subgroup E^{S_3} yields the table 4.15 again.

The tables show that $[S_2]^{S_3}$ has one orbit only, it is transitive, while $S_2^{S_3}$ has two orbits and E^{S_3} has four of them.

Other groups which are of special interest in graphical enumeration problems are certain subgroups which are induced by permutation groups on certain subsets of the power set of the set of symbols.

4.16 Def.: Let P denote a permutation group which acts on a finite set X.

(i) If for $1 \leq k \leq |X|$, $X^{(k)}$ denotes the set of subsets of order k of X, then

$$\delta_{(k)} : P \longrightarrow S_{X^{(k)}} : \pi \mapsto (\{x_1,\ldots,x_k\} \mapsto \{\pi(x_1),\ldots,\pi(x_k)\})$$

is a permutation representation of P (and faithful, if $k < |X|$). Its image

$$P^{(k)} := \delta_{(k)}[P]$$

is called <u>the k-sets group of</u> P.

(ii) The image of

$$\delta_k : P \longrightarrow S_{X^k} : \pi \mapsto ((x_1,\ldots,x_k) \mapsto (\pi(x_1),\ldots,\pi(x_k))),$$

i.e. the group

$$P^k := \delta_k[P]$$

is called <u>the k-sequences group of</u> P. δ_k is faithful.

(iii) The diagonal $\text{diag}X^k$ of X^k is an orbit of δ_k, so that the restriction of $\delta_k[P]$ to $X^k \backslash \text{diag}X^k$ is a permutation representation if $k > 1$ and $|X| > 1$, it is denoted by $\delta_{[k]}$:

$$\delta_{[k]} : \pi \mapsto \delta_k(\pi) | X^k \backslash \text{diag}X^k$$

Its image

$$P^{[k]} := \mathfrak{S}_{[k]}[P]$$

is called <u>the reduced k-sequences group</u> of P.

In particular the k-groups of symmetric groups are of interest.

In terms of representation theory they can be described as follows:

4.17 <u>$S_n^{(k)}$ is similar to the permutation group induced by S_n on the left cosets of $S_k \times S_{n-k}$ via left multiplication. (If $P \leq S_n$, then $P^{(k)}$ is similar to the restriction of this permutation representation of S_n to P.)</u>

The check is easy.

4.17 implies for the permutation character of these groups:

4.18 <u>$S_n^{(k)}$ (as representation of S_n) has $[k][n-k]$ as permutation character.</u>

The irreducible constituents of $[k][n-k]$ can be obtained by an application of the Littlewood-Richardson-rule (I. 4.51) which yields

4.19 $S_n^{(k)}$ (as representation of S_n) has the character

$\sum_{r=o}^{k} [n-r,r]$, if $k \leq n/2$.

An important example is that

4.20 $[n-2,2] + [n-1,1] + [n]$

is the character of $S_n^{(2)}$.

A typical problem of graphical enumeration is the following one: What is the number of types of graphs (without multiple edges and without loops, i.e. "Michigan-graphs") with n points?

The n points of such a graph form $\binom{n}{2}$ pairs of points so that the graph itself may be considered as a mapping φ from this set of $\binom{n}{2}$ pairs of points into a two element set, say $\{0,1\}$, where $\varphi(i) = 0$ means that the pair with number i is disconnected while $\varphi(j) = 1$ indicates that the pair with the number j is connected.

Two such graphs φ and ψ are said to be of the same type if and only if there is a $\pi \in S_n$ which by application to the n points yields ψ from φ, i.e. if and only if there is a $\pi \in S_n$ such that

$$\forall\ 1 \leq i \leq \binom{n}{2}\ (\varphi(\delta_{(2)}(\pi^{-1})(i)) = \psi(i)).$$

Hence the number of types of graphs with n points is equal to the number of orbits of

4.21 $$\{1_{S_2}\}^{S_n^{(2)}}$$

Let us consider the case when n := 3 for an example. Since $S_3^{(2)}$ is similar to S_3 we can use the table 4.15 which shows that E^{S_3} has four orbits so that there are 4 types of graphs with three points:

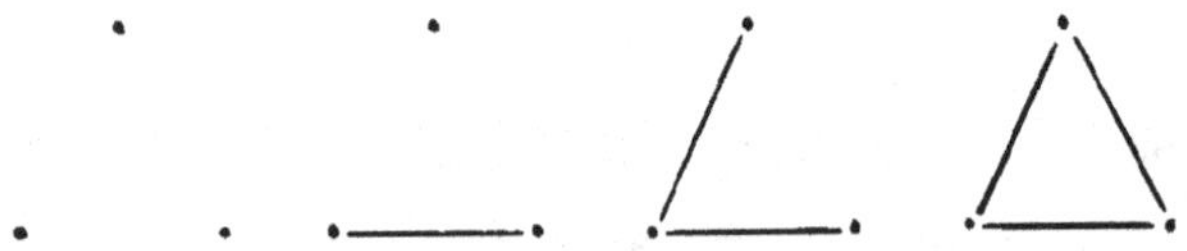

Fig. 2

If we consider the types of graphs up to complementation only (i.e. if we allow to substitute disconnected for connected and connected for disconnected), then we obtain the number of orbits of $S_2^{S_3^{(2)}}$ as solution. Hence 4.15 shows that there are just two types of graphs up to complementation which contain exactly four points. It is obvious that these are the following two:

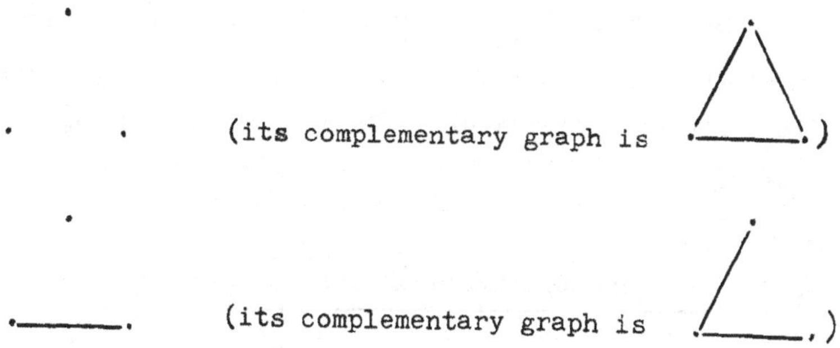

Fig. 3

In the next section we shall return to this problem since then we shall have additional tools.

Now one may ask for the number of orbits of $P^{(k)}$, where $P \leq S_n$, so that $P^{(k)} \leq S_n^{(k)}$.
Livingstone and Wagner gave a theorem on this which turned out to be very useful (cf. Livingstone/Wagner [1]), Robinson simplified the proof (Robinson [7]). The theorem reads as follows:

4.23 <u>If $P \leq S_n$ and $2 \leq k \leq n/2$, then $P^{(k)}$ has at least as many orbits as $P^{(k-1)}$.</u>

<u>Proof</u>: 4.17 yields that $[n-k][k] \downarrow P$ is the character of the permutation representation $P^{(k)}$ of P. Hence 4.19 yields for the number of orbits of $P^{(k)}$:

4.22 $$|P|^{-1}\sum_{\pi \in P} \chi^{[n-k][k]}(\pi) = |P|^{-1}\sum_{\pi \in P} \sum_{r=0}^{k} \zeta^{(n-r,r)}(\pi)$$

$$= |P|^{-1}\sum_{\pi \in P} \sum_{r=0}^{k-1} \zeta^{(n-r,r)}(\pi) + \underbrace{|P|^{-1}\sum_{\pi \in P} \zeta^{(n-k,k)}(\pi)}_{\geq 0}$$

$$\geq |P|^{-1}\sum_{\pi \in P} \sum_{r=0}^{k-1} \zeta^{(n-r,r)}(\pi) = \text{no. of orbits of } P^{(k-1)}.$$

q.e.d.

While the 2-sets group $S_n^{(2)}$ of S_n counts graphs of n points, the 2-sequences group S_n^2 of S_n counts relations on n points. The following obvious lemma yields the character of the k-sequences group:

<u>4.24</u> <u>S_n^k has (as representation of S_n) the character</u>

$$\overset{k}{\otimes} \vee S_n = \overset{k}{\otimes}([n] + [n-1,1]) = \sum_{r=0}^{k} \binom{k}{r} \overset{r}{\otimes}[n-1,1]$$

<u>(we put $\binom{k}{0} \overset{0}{\otimes}[n-1,1] := [n]$).</u>

For k := 2 we obtain the character of S_n^2 :

$$\overset{2}{\otimes}([n] + [n-1,1]) = [n] + 2[n-1,1] + \overset{2}{\otimes}[n-1,1].$$

Since $(\overset{2}{\otimes}[n-1,1], [n]) = ([n-1,1], [n-1,1]) = 1$, we obtain

<u>4.25</u> <u>S_n^2 has exactly two orbits.</u>

This follows also directly from the definition of S_n^2. One of these two orbits of S_n^2 consists of the pairs (i,i), $1 \leq i \leq n$. Thus $S_n^{[2]}$ possesses exactly one orbit:

<u>4.26</u> <u>$S_n^{[2]}$ is transitive.</u>

4.24 yields for the character of $S_n^{[k]}$:

<u>4.27</u> <u>$S_n^{[k]}$ has (as representation of S_n) the character</u>

$$\underline{(\overset{k}{\otimes} \vee S_n) - \vee S_n = \sum_{r=2}^{k} \binom{k}{r} \overset{r}{\otimes} [n-1,1].}$$

This yields

4.28 (i) The number of orbits of S_n^k is equal to

$$\sum_{r=0}^{k} \binom{k}{r} (\overset{r}{\otimes} [n-1,1], [n]).$$

(ii) The number of orbits of $S_n^{[k]}$ is equal to

$$\sum_{r=2}^{k} \binom{k}{r} (\overset{r}{\otimes} [n-1,1], [n]).$$

For the number of orbits of P^k and $P^{[k]}$, where $P \leq S_n$, we obtain

$$(\overset{k}{\otimes} \vee S_n \downarrow P, IP) = (\overset{k-1}{\otimes} \vee S_n \downarrow P, \vee S_n \downarrow P)$$

$$= (\overset{k-1}{\otimes} \vee S_n \downarrow P, [n] \downarrow P + [n-1,1] \downarrow P)$$

$$= (\overset{k-1}{\otimes} \vee S_n \downarrow P, [n] \downarrow P) + \underbrace{(\overset{k-1}{\otimes} \vee S_n \downarrow P, [n-1,1] \downarrow P)}_{\geq 0}$$

$$\geq (\overset{k-1}{\otimes} \vee S_n \downarrow P, IP).$$

Hence we have

4.29 (i) P^k has at least as many orbits as has P^{k-1},

(ii) $P^{[k]}$ has at least as many orbits as has $P^{[k-1]}$.

Let us consider the case n := 2 for an example.

We have

4.30 $$\overset{2}{\otimes}([n] + [n-1,1]) = [n] + 2\cdot[n-1,1] + \overset{2}{\otimes}[n-1,1]$$

$$= 2\cdot[n] + 3\cdot[n-1,1] + [n-2,2] + [n-2,1^2],$$

for the last equation confer Murnaghan [3] or use a character table.

This yields e.g. that S_n^2 has 2 orbits while $S_n^{[2]}$ is transitive.

If k := 3 we obtain for the number of orbits:

$$(\overset{3}{\otimes} \downarrow S_n,[n]) = \binom{3}{0}(\overset{0}{\otimes}[n-1,1],[n]) + \binom{3}{1}(\overset{1}{\otimes}[n-1,1],[n])$$

$$+ \binom{3}{2}(\overset{2}{\otimes}[n-1,1],[n]) + \binom{3}{3}(\overset{3}{\otimes}[n-1,1],[n])$$

$$= 1\cdot 1 + 3\cdot 0 + 3\cdot 1 + 1\cdot(\overset{3}{\otimes}[n-1,1],[n]).$$

Since (cf. Murnaghan [3]):

$$(\overset{3}{\otimes}[n-1,1],[n]) = (\overset{2}{\otimes}[n-1,1],[n]),$$

we obtain that S_n^3 has exactly 5 orbits (and hence that $S_n^{[3]}$ has exactly 4 of them).

The orbit of the k-tuple $(i_1,\dots,i_k)$ under S_n^k is obviously characterized by the indices μ of the coordinates i_μ of $(i_1,\dots,i_k)$ which are equal. Hence there are as many orbits of S_n^k as there are partitions of the set $\{1,\dots,k\}$. I.e. the number of orbits of S_n^k is just the <u>Bell number</u> B_K (cf. Comtet [1]). This yields a characterization of Bell numbers in terms of representation theory:

4.31 $$\forall n \geq k \ \left(B_k = \left(\overset{k}{\otimes} \vee S_n, [n]\right) = \sum_{r=0}^{k} \binom{k}{r} \left(\overset{r}{\otimes} [n-1,1], [n]\right)\right).$$

(Notice that B_k and hence also $(\overset{k}{\otimes} \vee S_n, [n])$ is independent of n!)

The first values of B_k are (cf. Comtet [1], p. 212):

4.32

k	1	2	3	4	5	6	7
B_k	1	2	5	15	52	203	877

There are many results known about Bell numbers, recursion formulae, generating functions etc. (cf. Comtet [1]), 4.31 shows how these may be applied to representation theory and conversely.

We have discussed $S_n^{(k)}$, S_n^k and $S_n^{[k]}$ a little bit. Besides these three there is a fourth representation of S_n which is of some importance in combinatorics. It is the representation, the image of which is induced by S_n on the k-tuples of <u>distinct</u> elements of $\{1,\dots,n\}$. Let us denote this group by

$S_n^{\langle k\rangle}$

The following is obvious:

<u>4.33</u> <u>$S_n^{\langle k\rangle}$ has (as representation of S_n) the character</u>

$$\underline{IS_{n-k} \uparrow S_n = [n-k] \uparrow S_n.}$$

Since

4.34 $$[n-k] \uparrow S_n = [n-k] \underbrace{[1] \ldots [1]}_{k \text{ factors}},$$

the Littlewood-Richardson rule yields that each $[\alpha]$, where $\alpha \vdash n$ and $\alpha_1 \geq n-k$, occurs under the irreducible constituents of $[n-k] \uparrow S_n$.

Since

$$[n-k] \underbrace{[1] \ldots [1]}_{k} = ([n-k][1]) \underbrace{[1] \ldots [1]}_{k-1}$$

$$= ([n-k+1] + [n-k,1]) \underbrace{[1] \ldots [1]}_{k-1}$$

$$= [n-k+1] \underbrace{[1] \ldots [1]}_{k-1} + \ldots,$$

the following holds for $P^{\langle k\rangle}$, the subgroup of $S_n^{\langle k\rangle}$ corresponding to $P \leq S_n$:

4.35 $P^{\langle k\rangle}$ has at least as many orbits as has $P^{\langle k-1\rangle}$

There is a famous theorem of Livingstone and Wagner (Livingstone/Wagner [1], theorem 2) which is very difficult to prove and which reads as follows:

4.36 If $2 \leq k \leq \frac{n}{2}$ and $P \leq S_n$ and $P^{(k)}$ transitive, then

a) $P^{\langle k-1\rangle}$ is transitive (so that P is (k-1)-fold transitive),

b) if also $k \geq 5$, then $P^{\langle k\rangle}$ is transitive (so that P is even k-fold transitive).

In the light of 4.19 we derive from 4.36 the surprising result which is equivalent to theorem 4.36 a) :

4.37 If $P \leq S_n$, $2 \leq k \leq \frac{n}{2}$, and if

$$\forall\, 0 < r \leq k \; (([n-r,r] \downarrow P, IP) = 0),$$

then even

$$\forall\, \alpha \vdash n \; (n-k+1 \leq \alpha_1 < n \Rightarrow ([\alpha] \downarrow P, IP) = 0).$$

4.36 b) can be formulated similarly.

We leave it here and shall return to multiply transitivity at the end of the last section.

5. Enumeration of functions by weight

In the preceding section we discussed the enumeration of the orbits of E^H, G^H and $[G]^H$. This done we are in a position to solve problems like the introductory one where we asked for the number of necklaces with five beads in two colours. The solution of this introductory example is 8 as it is indicated in Fig. 1.

But Fig. 1 yields much more than this solution 8 only. In fact Fig. 1 shows a representative for each one of these 8 classes of necklaces which are the orbits of E^{D_5} on the set $\{1,2\}^{\{1,2,3,4,5\}}$. Part of this information given by Fig. 1 is how many necklaces there are which contain exactly 3 beads of colour •. In fact a somewhat incomplete description of Fig. 1 could read as follows:

1 necklace with 5 beads of colour • ,

1 necklace with 4 beads of colour • and 1 bead of colour ∘ ,

2 necklaces with 3 beads of colour • and 2 beads of colour ∘ ,

2 necklaces with 2 beads of colour • and 3 beads of colour ∘ ,

1 necklace with 1 bead of colour • and 4 beads of colour ∘ ,

1 necklace with 5 beads of colour ∘ .

In order to express this description of Fig. 1 in terms of elements of a mathematical structure we express the colour • by the variable x and the colour ∘ by the variable y of the polynomial ring $\mathbb{Q}[x,y]$. By indicating the number of beads in a certain colour by the exponent of the corresponding variable, we may express the preceding description of Fig. 1 by the following element of $\mathbb{Q}[x,y]$:

5.1 $$x^5 + x^4y + 2x^3y^2 + 2x^2y^3 + xy^4 + y^5.$$

This element of $\mathbb{Q}[x,y]$ "generates" the solution of the necklace problem. The sum of all its coefficients is the number of orbits and the coefficient of x^ay^b yields the number of orbits which consist of a beads in colour • and b beads in colour ∘ .

This shows that it makes sense to ask for a method to produce generating functions, which describe the orbits of $[G]^H$, G^H and E^H similarly.

It will turn out that a method can be described which yields such a polynomial, which is called <u>the pattern inventory</u> (since it inventories the patterns) from <u>the store enumerator</u>, which is again a polynomial but a generating function for M, the range of the functions considered (x+y for example describes the store {•, ∘} of the above example). In fact we shall derive a result which shows how 5.1 can be obtained from x+y by a well defined manipulation within $\mathbb{Q}[x,y]$. Again we discuss the problem with symmetry group $[G]^H$ first in order to get results corresponding to G^H and E^H as corollaries.

Given permutation groups $G \leq S_m$, $H \leq S_n$ define $[G]^H$ on M^N.

We assume that $w\colon M \to K$, the store enumerator, is a given function from M into a field K of characteristic 0 which is constant on the orbits of G (w can be suitably chosen for the problem considered).

w defines a <u>weight</u> $w^*(\varphi)$ for each $\varphi \in M^N$ by

5.2 $$w^*(\varphi) := \prod_{i \in N} w(\varphi(i)).$$

Since w is constant on the orbits of G, the following holds:

5.3 w^* is constant on the orbits of $[G]^H$, G^H, E^H.

The check is very easy.

Denoting by $\omega_1,\ldots,\omega_r$ the orbits of $[G]^H$ and by

$$w_i := w^*(\varphi),\quad \varphi \in \omega_i,$$

the value of w^* on ω_i, $1 \leq i \leq r$, the expression

5.4 $$\sum_{i=1}^{r} w_i \in K$$

gives us some information on the orbits of $[G]^H$ on M. In fact if for a given enumeration problem concerning $[G]^H$ the store enumerator w was carefully chosen, 5.4 is sometimes the desired solution of the enumeration problem in question.

Hence we would like to evaluate 5.4.

In order to do this we consider the representation $\widetilde{\#^n} \, \nu$ of $G \wr H$ which acts on $V := \overset{n}{\otimes} K^m$.

If $K^m = \langle\langle e_1,\ldots,e_m\rangle\rangle$, then (cf. the proof of 4.5)

$$\{e_\varphi := e_{\varphi(1)} \otimes \cdots \otimes e_{\varphi(n)} \mid \varphi \in M^N\}$$

is a basis of V. $\widetilde{\#^n} \, \nu$ induces a permutation group acting on this basis which is similar to $[G]^H$ since

$$(f;\pi)\, e_\varphi = \ldots \otimes e_{f(i)\varphi(\pi^{-1}(i))} \otimes \ldots \; .$$

The subspaces V_i which are generated by the orbits, i.e.

$$V_i := \langle e_\varphi \mid \varphi \in \omega_i \rangle,$$

are obviously invariant so that

5.5 $$V = \bigoplus_{i=1}^{r} V_i$$

is a decomposition of V into invariant subspaces. The representation afforded by V_i contains the identity representation exactly once since $[G]^H$ acts transitively on its orbits. Let us denote this subspace of V_i which affords the identity representation by V_{i1}. We have (recall the proof of Burnside's lemma):

$$V_{i1} = \langle \sum_{\varphi \in \omega_i} e_\varphi \rangle, \ 1 \leq i \leq r.$$

The centrally primitive idempotent

5.6 $$e := \frac{1}{|G|^n |H|} \sum_{(f;\pi) \in G \wr H} (f;\pi)$$

projects V onto the subspace

$$\bigoplus_{i=1}^{r} V_{i1} \leq V$$

so that its trace is r, the number of orbits.

In order to evaluate 5.4 we need a slight change only. Instead of the operator 5.6 we consider the <u>weighted operator</u>

5.7 $$e_W := \frac{1}{|G|^n |H|} \sum_{(f;\pi) \in G \wr H} W \circ (f;\pi) = W \circ e.$$

where W is the linear transformation which multiplies each basis vector by the weight of its orbit:

5.8 $$W : V \to V : e_\varphi \mapsto w^*(\varphi) \cdot e_\varphi$$

Gathering up we obtain (Lehmann [1]):

5.9 If w: $M \to K$ is a function from M into a field K of characteristic 0 which is constant on the orbits of G and if $w^* : M^N \to K$ denotes the corresponding weight function defined by

$$w^*(\varphi) := \prod_{i\in N} w(\varphi(i)),$$

then w^* is constant on each orbit of $[G]^H$. If w_i denotes the value of w^* on the i-th orbit of $[G]^H$, $1 \leq i \leq r$, then the pattern inventory satisfies the following equation:

$$\sum_{i=1}^{r} w_i = \frac{1}{|G|^n|H|} \sum_{(f;\pi)\in G\wr H} \operatorname{trace}(W \circ (f;\pi)),$$

where W denotes the linear transformation defined by

$$W: \overset{n}{\otimes} K^m \to \overset{n}{\otimes} K^m: e_\varphi \mapsto w^*(\varphi)e_\varphi$$

We are left with the question for the trace of $W \circ (f;\pi)$. In order to provide an answer we proceed as in the proofs of 2.5 and 2.6 where the trace of $(f;\pi)$ was derived.

Putting

$$U := \begin{pmatrix} w(1) & & 0 \\ & \ddots & \\ 0 & & w(m) \end{pmatrix}$$

and denoting by

$$\mathbb{D}(g) = (d_{ik}(g))$$

the matrix representation associated with the natural representation νG on K^m with respect to the basis $\{e_1,\ldots,e_m\}$ we have:

$$W\circ(f;(1\ldots n)) \underset{\nu}{\otimes} e_{i_\nu} = W(\sum_{i,\ldots,k} d_{ii_n}(f(1))\ldots d_{ki_{n-1}}(f(n))e_i \otimes\ldots\otimes e_k)$$

$$= \sum_{i,\ldots,k} w(i)d_{ii_n}(f(1))\ldots w(k)d_{ki_{n-1}}(f(n))\ e_i\otimes\ldots\otimes e_k.$$

This yields for the trace:

$$tr(W\circ(f;(1\ldots n))) = \sum_{i_1,\ldots,i_n} w(i_1)d_{i_1i_n}(f(1))\ldots w(i_2)d_{i_2i_1}(f(2))$$

$$= tr(U\cdot\mathbb{D}(f(1))\cdot U\cdot\mathbb{D}(f(n))\ldots U\cdot\mathbb{D}(f(2)))$$

$$= tr(U^n\cdot\mathbb{D}(f(1)\cdot f(n)\ \ldots\ f(2))).$$

Since $f(1)\cdot f(n)\ldots f(2)$ is the cycle product associated with $(1\ldots n)$ with respect to f, for a general element we obtain similarly:

<u>5.10</u> $$tr(W\circ(f;\pi)) = \prod_{\nu=1}^{c(\pi)} tr(U^{k_\nu}\cdot\mathbb{D}(g_\nu(f;\pi))) = \prod_{\nu=1}^{c(\pi)} \sum_{j\in Fix g_\nu(f;\pi)} w(j)^{k_\nu},$$

where k_ν denotes the length of the ν-th cyclic factor π_ν of π:

$$\pi_\nu = (j_\nu\ \pi(j_\nu)\ldots\pi^{k_\nu-1}(j_\nu)),$$

and where $g_\nu(f;\pi)$ is the corresponding cycle product:

$$g_\nu(f;\pi) = f(j_\nu)f(\pi^{-1}(j_\nu))\dots f(\pi^{-k_\nu+1}(j_\nu)).$$

This altogether yields the desired enumeration theorem (Lehmann [1]):

5.11 ("Exponentiation group enumeration theorem, weighted form")
Under the assumption of 5.9 we have for the pattern inventory of $[G]^H$:

$$\sum_{i=1}^{r} w_i = \frac{1}{|G|^n|H|} \sum_{(f;\pi)\in G\wr H} \prod_{\nu=1}^{c(\pi)} \sum_{j\in \mathrm{Fix}(g_\nu(f;\pi))} w(j)^{k_\nu}$$

Restricting to G^H yields (de Bruijn [2], Harary/Palmer [2]):

5.12 ("Power group enumeration theorem, weighted form")
Under the assumption of 5.9 we have for the pattern inventory of G^H:

$$\sum_{i=1}^{s} w_i = \frac{1}{|G||H|} \sum_{(g,\pi)\in G\times H} \prod_{k=1}^{n} \Big(\sum_{j\in \mathrm{Fix}(g^k)} w(j)^k \Big)^{a_k(\pi)}$$

A further restriction down to E^H yields (Pólya [1]):

5.13 The pattern inventory of E^H is

$$\sum_{i=1}^{t} w_i = \frac{1}{|H|} \sum_{\pi\in H} \prod_{k=1}^{n} \Big(\sum_{j=1}^{m} w(j)^k \Big)^{a_k(\pi)}$$

(In this case every function $w\colon M \to K$ satisfies the assumption of 5.9 since $G := \{1_{S_m}\}$.)

The following definition enables us to systematize our approach:

5.14 Def.: If P is a permutation group acting on a finite set X, i.e. $P \leq S_X$, then the following element of the polynomial ring $\mathbb{Q}[x_1,\ldots,x_{|X|}]$ is called <u>the cycle-index of P</u>:

$$\mathrm{Cyc}(P) := \frac{1}{|P|}\sum_{p\in P}\prod_{k=1}^{|X|} x_k^{a_k(p)}.$$

We shall sometimes display the variables, writing

$$\mathrm{Cyc}(P;x_1,\ldots,x_{|X|})$$

instead of $\mathrm{Cyc}(P)$.

Using this notation we define what we understand by <u>Pólya-insertion</u> of the polynomial $f(x,y,z,\ldots)$

$$\mathrm{Cyc}(P|f(x,y,z,\ldots)) := \mathrm{Cyc}(P;f(x,y,z,\ldots),\ldots,f(x^{|X|},y^{|X|},z^{|X|},\ldots))$$

$$= \frac{1}{|P|}\sum_{p\in P}\prod_{k=1}^{|X|} f(x^k,y^k,z^k,\ldots)^{a_k(p)}.$$

This yields a simple expression for 5.13:

5.15 <u>The pattern inventory of E^H is obtained from Cyc(H) by Pólya-insertion of the store enumerator, i.e. it is equal to</u>

$$\underline{\mathrm{Cyc}(H|\sum_{j=1}^{m} w(j))}$$

Let us list some examples of cycle-indices of special groups:

5.16 (i) $\mathrm{Cyc}(\{1_{S_m}\}) = x_1^m$,

(ii) $\mathrm{Cyc}(C_n := \langle(1\ldots n)\rangle) = \frac{1}{n}\sum_{i|n} \Phi(i) x_i^{n/i}$

(where Φ denotes the Euler function, i.e.

$\Phi(i) := |\{k \in \mathbb{N} | k \leq i \wedge (k,i) = 1\}|$),

(iii) $\mathrm{Cyc}(S_n) = \sum_{a \vdash\!\dashv n} \prod_{k=1}^{n} \frac{1}{a_k!} \left(\frac{x_k}{k}\right)^{a_k}$

(iv) $\mathrm{Cyc}(A_n) = \sum_{a \vdash\!\dashv n} (1+(-1)^{a_2+a_4+\ldots}) \prod_{k=1}^{n} \frac{1}{a_k!} \left(\frac{x_k}{k}\right)^{a_k}$,

(v) $\mathrm{Cyc}(D_n) = \frac{1}{2}\,\mathrm{Cyc}(C_n) + \begin{cases} \frac{1}{2} x_1 x_2^{(n-1)/2}, & \text{if } n \text{ is odd} \\ \frac{1}{4}(x_2^{n/2} + x_1^2 \cdot x_2^{(n-2)/2}), & \text{if } n \text{ is even.} \end{cases}$

(vi) The cycle-index of the regular representation RG of a finite group G:

$$\mathrm{Cyc}(RG) = \frac{1}{|G|} \sum_{k||G|} \omega(k)\, x_k^{|G|/k} = \frac{1}{|G|} \sum_{g \in G} x_{|\langle g\rangle|}^{|G|/|\langle g\rangle|},$$

where $\omega(k)$ denotes the number of elements in G with order k.

5.16 (iv), (v) yield e.g.

$$\mathrm{Cyc}(D_5) = \frac{1}{2}\,\mathrm{Cyc}(C_5) + \frac{1}{2}\,x_1 \cdot x_2^2$$

$$= \frac{1}{10}\,(x_1^5 + 5x_1x_2^2 + 4x_5)$$

so that we obtain for the cycle inventory of E^{D_5} (apply 5.15 to $w\colon \{1,2\} \to \mathbb{Q}[x,y] : \begin{smallmatrix} 1 \mapsto x \\ 2 \mapsto y \end{smallmatrix}$):

$$\mathrm{Cyc}(D_5|x+y) = \frac{1}{10}\,((x+y)^5 + 5(x+y)(x^2+y^2)^2 + 4(x^5 + y^5))$$

$$= x^5 + x^4y + 2x^3y^2 + 2x^2y^3 + xy^4 + y^5.$$

This is the polynomial 5.1 which describes Fig. 1, the solution of the introductory enumeration problem. Notice that it is obtained by Pólya-inserting x+y into the cycle-index of D_5. This process can be generalized by inserting $\sigma_{1,t} = x_1 + \ldots + x_t$ (t an arbitrary number which may be suitably chosen for the enumeration in question). The resulting polynomial is Redfield's group reduction function:

5.17 $\mathrm{Grf}(P,t) := \mathrm{Cyc}(P|\sigma_{1,t}).$

$$= \frac{1}{|P|} \sum_{p \in P} \prod_{k=1}^{|X|} \sigma_{k,t}^{a_k(p)}$$

Comparing this with the definition of Schur-function we obtain that

5.18 $\qquad \mathrm{Grf}(S_n,m) = \{n\},$

a basic result which shows how S-function enter the theory of enumeration under group action.

In 5.18 only the special S-functions $\{n\}$ occur. The general case $\{\alpha\}$ occurs if we consider a generalization of cycle-index which is defined as follows:

5.19 Def.: If P denotes a permutation group acting on a finite set X and if F is an ordinary representation of P with character χ^F, then the polynomial

$$\mathrm{Cyc}(P,F) := \frac{1}{|P|}\sum_{p\in P}\chi^F(p)\prod_{k=1}^{|X|} x_k^{a_k(p)}$$

is called the generalized cycle-index of P with respect to F.

Hence

5.20 $$\mathrm{Cyc}(P) = \mathrm{Cyc}(P,I),$$

if I denotes the identity representation of P. We obtain furthermore:

5.21 $$\alpha \vdash n \Rightarrow \{\alpha\} = \mathrm{Cyc}(S_n,[\alpha]|\sigma_{1,n})$$

We prove now an interesting lemma of Foulkes (Foulkes [1]):

5.22 $$P \leq S_n \Rightarrow \mathrm{Cyc}(P,F) = \mathrm{Cyc}(S_n, F\uparrow S_n)$$

Proof:

$$\mathrm{Cyc}(S_n,FP\uparrow S_n) = \frac{1}{n!}\sum_{\pi\in S_n}\chi^{F\uparrow S_n}(\pi)\prod_{k=1}^{n} x_k^{a_k(\pi)}$$

$$= \frac{1}{n!}\sum_{\pi\in S_n}\frac{n!}{|P||C^{S_n}(\pi)|}\sum_{p\in C^{S_n}(\pi)\cap P}\chi^F(p)\prod_{k=1}^{n} x_k^{a_k(\pi)}$$

$$= \frac{1}{|P|} \sum_{p \in P} \chi^F(p) \prod_{k=1}^{n} x_k^{a_k(p)} = \mathrm{Cyc}(P,F).$$

q.e.d.

A corollary is

<u>5.23</u> $P \leq S_n \Rightarrow \mathrm{Cyc}(P) = \mathrm{Cyc}(S_n, IP \uparrow S_n)$.

This shows how S-functions can be used in enumeration theory.

5.23 implies for example that

$$\begin{aligned} \mathrm{Cyc}(A_n) &= \mathrm{Cyc}(S_n, IA_n \uparrow S_n) \\ &= \mathrm{Cyc}(S_n, [n]+[1^n]) \\ &= \mathrm{Cyc}(S_n, [n]) + \mathrm{Cyc}(S_n, [1^n]), \end{aligned}$$

so that

$$\mathrm{Grf}(A_n, n) = \{n\} + \{1^n\}.$$

Applications of this fact will be discussed later. Let us return to 5.11, 5.12 and 5.13. In the case when we take for w the trivial weight function

$$w : M \longrightarrow K : m \longmapsto 1_K,$$

then we obtain the number of orbits of $[G]^H$, G^H, E^H respectively. Let us consider the graph theoretical example of section 4 under this new aspect. We saw that the number of types of graphs with n points is equal to the number of orbits of $E^{S_n^{(2)}}$, 5.13 yields now:

<u>5.24</u> <u>The number of types of graphs with n points is equal to</u>
<u>$\mathrm{Cyc}(S_n^{(2)}\,|\,2) = \mathrm{Cyc}(S_n^{(2)};2,2,\ldots)$.</u>

We saw furthermore that the number of types of graphs with n points up to complementation is equal to the number of orbits of $S_2^{S_n^{(2)}}$. Hence by 5.12 it is equal to

$$\frac{1}{2n!}\Big(\sum_{\pi\in S_n^{(2)}} \prod_{k=1}^{\binom{n}{2}} 2^{a_k(\pi)} + \sum_{\pi\in S_n^{(2)}} \prod_{\substack{k=1\\ 2|k}}^{\binom{n}{2}} 2^{a_k(\pi)}\Big)$$

$$= \frac{1}{2}\,(\mathrm{Cyc}(S_n^{(2)};\,2,2,\ldots) + \mathrm{Cyc}(S_n^{(2)};\,0,2,0,2,\ldots))$$

Since this consideration up to complementation divides the types into pairs of types and the self-complementary graphs, we obtain furthermore the well-known result:

5.25 The number of self-complementary graphs with n points is equal to

$$\mathrm{Cyc}(S_n^{(2)};\ 0,\ 2,\ 0,\ 2,\ldots)$$

For twice the number of types of graphs up to complementation is equal to the number of types of graphs plus the number of types of self-complementary graphs (with n points in each case), so that we need only apply 5.24.

If for example $n := 3$, then we obtain for the number of self-complementary graphs with three points:

$$\mathrm{Cyc}(S_3^{(2)};\ 0,2,0) = \mathrm{Cyc}(S_3;\ 0,2,0)$$

$$= \frac{1}{6}\,(x_1^3 + 3x_1x_2 + 2x_3)_{x_1=x_3=0,\ x_2=2} = 0.$$

Hence there is no such graph as Fig. 2 shows as well.
A formula for the number of self-complementary m-placed relations can be found in Wille [1].

6. Some cycle-indices

I would like first to discuss how cycle-indices can be evaluated in some special cases.
Let us first consider how the cycle-index of certain products of groups can be evaluated once the cycle-indices of the factors are known.

6.1 Def.: If P, Q denote permutation groups on finite sets X, Y, where $X \cap Y = \emptyset$, then we call

(i) <u>the direct sum</u> $P \oplus Q$ <u>of</u> P <u>and</u> Q the permutation group with underlying set $P \times Q$ which acts on $X \dot{\cup} Y$ as follows:

$$(p,q)(i) := \begin{cases} p(i), & \text{if } i \in X \\ q(i), & \text{if } i \in Y \end{cases}$$

(ii) <u>the direct product</u> $P \otimes Q$ <u>of</u> P <u>and</u> Q the permutation group with underlying set $P \times Q$ which acts on $X \times Y$ as follows:

$$(p,q)(i,k) := (p(i),q(k)).$$

It is obvious that the following holds:

6.2 $\quad \underline{Cyc(P \oplus Q) = Cyc(P) \cdot Cyc(Q).}$

In order to determine the cycle-index of the direct product $P \otimes Q$ we remark that an i-cycle of $p \in P$ and a k-cycle of $q \in Q$ together yield just gcd(i,k) cycles of length lcm(i,k). This yields

$$\underline{\underline{6.3}} \quad Cyc(P \otimes Q) = \frac{1}{|P||Q|} \sum_{(p,q)} \prod_{i,k=1}^{|X|,|Y|} x_{\mathrm{lcm}(i,k)}^{\mathrm{gcd}(i,k)a_i(p)a_k(q)} .$$

Other products of permutation groups which are of importance in combinatorics are e.g. the groups E^H, G^H and $[G]^H$ which were mentioned in the preceeding sections and the permutation representation $\varphi[G\sim H]$ which was introduced in section 1, it is usually denoted H[G] in combinatorics:

6.4 $\quad H[G] := \varphi[G\sim H].$

Let us call this group <u>the composition</u> of G and H. Combinatorists called it the wreath product since they considered wreath products of permutation groups mainly and $\varphi[G\sim H]$ then is the natural representation of $G\sim H$ in a sense.

Besides these two permutation representations H[G] and $[G]^H$ of the wreath product $G\sim H$ of $G \leq S_m$ and $H \leq S_n$ there is a third one which occurs as symmetry group of combinatorial structures.

In order to define this group we consider the set W of all the nxm-matrices in which each row contains all the elements of $\{1,\ldots,m\}$.

We notice that

$$|W| = m!^n,$$

so that the permutation representation

6.5 $\sigma : S_m \sim S_n \to S_W : (f;\pi) \mapsto ((a_{ik}) \mapsto (f(i)a_{\pi^{-1}(i),k}))$

is of dimension $m!^n$.

An easy check shows that $\sigma[S_m \sim S_n]$ is similar to the permutation group induced by $\overset{n}{\#} RS_m$ (RS_m the regular representation of S_m) on the natural basis of the representation space. Let us denote the subgroup $\sigma[G \sim H]$ by $[G]_H$:

6.6 $$[G]_H := \sigma[G \sim H].$$

An application of 2.6 yields for its permutation character:

6.7 $$a_1(\sigma(f;\pi)) = \prod_{\nu=1}^{c(\pi)} \chi^R(g_\nu(f;\pi)) = \begin{cases} m!^{\,c(\pi)}, & \text{if } \sum_k a_{1k}(f;\pi) = c(\pi) \\ 0 & \text{, elsewhere} \end{cases}$$

In order to define a further interesting permutation representation we introduce an equivalence relation on W.
Two elements (a_{ik}), (b_{ik}) of W are called equivalent if an only if (b_{ik}) arises from (a_{ik}) by a permutation of the columns of (a_{ik}).
In other words:

$$(a_{ik}) \sim (b_{ik}) :\Longleftrightarrow \exists \rho \in S_m \, ((a_{ik}) = (b_{i\rho(k)})).$$

Denoting the class of (a_{ik}) under "$\sim$" by $[(a_{ik})]_\sim$, we obtain from σ a further permutation representation of $S_m \sim S_n$:

6.8 $\tau : S_m \sim S_n \to S_{W/\sim} : (f;\pi) \mapsto ([(a_{ik})]_\sim \mapsto [\sigma(f;\pi)(a_{ik})]_\sim)$.

In terms of representation theory it can be characterized as follows:

<u>6.9</u> <u>$\tau[S_m \sim S_n]$ is similar to the permutation group induced by (the left multiplications of elements of) $S_m \sim S_n$ on the left cosets of $\mathrm{diag}S_m{*}S_n'$.</u>

The check is easy. It implies for the character:

<u>6.10</u> <u>The permutation character of $\tau[S_m \sim S_n]$ is the character of $I\ \mathrm{diag}S_m{*}S_n' \uparrow S_m \sim S_n$, i.e.</u>

$$a_1(\tau(f;\pi)) = m!^{n-1} \frac{|C^{S_m \sim S_n}(f;\pi) \cap \mathrm{diag}S_m{*}S_n'|}{|C^{S_m \sim S_n}(f;\pi)|}$$

The subgroup $\tau[G \sim H]$ (for $G \leq S_m$, $H \leq S_n$) of $\tau[S_m \sim S_n]$ was introduced by E.M. Palmer and R.W. Robinson (Palmer/Robinson [1],[2]), denoted by $[H;G]$ and called <u>the matrix group of</u> G <u>and</u> H:

6.11 $$[H;G] := \tau[G \sim H] \leq \tau[S_m \sim S_n].$$

6.10 implies

<u>6.12</u> <u>The permutation character of the matrix group $[H;G]$ is the character of $I\ \mathrm{diag}S_m{*}S_n' \uparrow S_m \sim S_n \downarrow G \sim H$.</u>

Having obtained the permutation characters of $H[G]$, $[G]^H$, $[G]_H$ and $[H;G]$ we ask for the cycle-indices. In order to evaluate these we consider the equation which holds for the elements $a_1(p)$ of the cycle type $(a_1(p),\ldots,a_{|X|}(p))$ of an element $p \in P$ acting on a finite set X:

6.13 $\forall r \in \mathbb{N} \; (a_1(p^r) = \sum_{s|r} s a_s(p)).$

An application of a Moebius-inversion to this equation yields

6.14 $a_s(p) = \frac{1}{s} \sum_{r|s} \mu(\frac{s}{r}) \, a_1(p^r).$

This together with the preceding results on the permutation characters yields the desired cycle-indices:

6.15 ("Cycle indices of H[G], $[G]^H$, $[G]_H$, [H;G]")

(i) The cycle-index of the composition H[G] is equal to

$$\frac{1}{|G|^n|H|} \sum_{(f;\pi)\in G\wr H} \prod_{k=1}^{m\cdot n} x_k^{a_k(\varphi(f;\pi))},$$

where

$$a_k(\varphi(f;\pi)) = \frac{1}{k} \sum_{i|k} \mu(\frac{k}{i}) \sum_{j\in Fix(\pi^i)} a_1(f(j))$$

(ii) The cycle-index of the exponentiation $[G]^H$ is equal to

$$\frac{1}{|G|^n|H|} \sum_{(f;\pi)\in G\wr H} \prod_{k=1}^{m^n} x_k^{a_k(\rho(f;\pi))},$$

where

$$a_k(\rho(f;\pi)) = \frac{1}{k} \sum_{i|k} \mu(\frac{k}{i}) \prod_{\nu=1}^{c(\pi)} a_1(g_\nu(f\ldots f_{\pi^{i-1}};\pi^i))$$

(iii) The cycle-index of $[G]_H$ is equal to

$$\frac{1}{|G|^n|H|}\sum_{(f;\pi)\in G\sim H}\prod_{k=1}^{m!^n} x_k^{a_k(\sigma(f;\pi))},$$

where

$$a_k(\sigma(f;\pi)) = \frac{1}{k}\sum_{i|k}\mu\left(\frac{k}{i}\right)\cdot m!^{c(\pi^i)}\cdot \delta_{c(\pi^i),\Sigma a_{1j}(f\ldots f_{\pi^{i-1}};\pi^i)}$$

(iv) The cycle-index of $[H;G]$ is equal to

$$\frac{1}{|G|^n|H|}\sum_{(f;\pi)\in G\sim H}\prod_{k=1}^{m!^{n-1}} x_k^{a_k(\tau(f;\pi))},$$

where

$$a_k(\tau(f;\pi)) = \frac{1}{k}\sum_{i|k}\mu\left(\frac{k}{i}\right)m!^{n-1}\frac{|C^{S_m\sim S_n}((f;\pi)^i)\cap \operatorname{diag}S_m*S_n'|}{|C^{S_m\sim S_n}((f;\pi)^i)|}$$

There are other expressions obtainable for the cycle-indices of $H[G]$ (see Pólya [1]) and $[G]^H$ (see Palmer/Robinson [1]) which are easier to handle since they express the cycle-index of these products in terms of the cycle-indices of the factors. But I preferred to stress the fact that a knowledge of the permutation character is sufficient.

The method how we obtained the cycle-index of $[H;G]$ points to a more general situation. If Q is a subgroup of a finite group P, then P induces a finite permutation group on the left cosets of Q in P. Denoting this permutation group by P/Q, we obtain

6.16 ("Exterior cycle-index")

$$\mathrm{Cyc}(P/Q) = \frac{1}{|P|} \sum_{p \in P} \prod_{k=1}^{|P:Q|} x_k^{a_k(p)}$$

where

$$a_k(p) := \frac{1}{k} \sum_{i|k} \mu\left(\frac{k}{i}\right) |P:Q| \frac{|C^P(p^i) \cap Q|}{|C^P(p^i)|}.$$

This generalizes a result of de Bruijn (de Bruijn [3]). It holds since the permutation character of P/Q is the character $IQ \uparrow P$ of P, and

$$\chi^{IQ\uparrow P}(p) = |P:Q| \frac{|C^P(p) \cap Q|}{|C^P(p)|}$$

as it is well known.

Further expressions for the cycle-indices of H[G], $[G]^H$, $[G]_H$ and [H;G] can be obtained by an application of 5.22 which says that

$$\mathrm{Cyc}(P) = \mathrm{Cyc}(S_n, IP \uparrow S_n).$$

Hence e.g.

6.17 $$\mathrm{Cyc}\,(S_n[S_m]) = \mathrm{Cyc}(S_{mn}, (m;n) \uparrow S_{mn})$$
$$= \mathrm{Cyc}(S_{mn}, [m] \odot [n])$$
$$= \frac{1}{(mn)!} \sum_{\pi \in S_{mn}} \chi^{[m] \odot [n]}(\pi) \prod_{k=1}^{mn} x_k^{a_k(\pi)}$$

This implies that

$$\mathrm{Grf}(S_n[S_m]) = \mathrm{Cyc}(S_n[S_m]\,|\,\sigma_{1,mn}) = \frac{1}{(mn)!}\sum_{\pi\in S_{mn}} \chi^{[m]\odot[n]}(\pi)\prod_{k=1}^{mn}\sigma_{k,mn}^{a_k(\pi)}.$$

In the same way we obtain:

<u>6.18</u> $$\mathrm{Cyc}(S_n[S_m],(\alpha;\beta)\,|\,\sigma_{1,mn}) = \frac{1}{(mn)!}\sum_{\pi\in S_{mn}} \chi^{[\alpha]\odot[\beta]}(\pi)\prod_{k=1}^{mn}\sigma_{k,mn}^{a_k(\pi)}$$

Hence tables concerning decompositions of symmetrized outer products $[\alpha]\odot[\beta]$ can be used in order ot obtain generalized cycle indices of compositions $S_n[S_m]$.

E.g.

$$[2]\odot[4] = [8] + [6,2] + [4^2] + [42^2] + [2^4]$$

This implies

$$\mathrm{Grf}(S_4[S_2]) = \{8\} + \{6,2\} + \{4^2\} + \{42^2\} + \{2^4\}$$

For the more general case 6.18 a (given) decomposition of $[\alpha]\odot[\beta]$ yields $\mathrm{Cyc}(S_n[S_m],(\alpha;\beta)\,|\,\sigma_{1,mn})$ as a sum of s-functions $\{\gamma\}$, $\gamma\vdash mn$. This sum is uniquely determined since the function $\mathrm{Cyc}(S_n[S_m],(\alpha;\beta)\,|\,\sigma_{1,mn})$ is symmetric and homogeneous of weight mn and the S-functions $\{\gamma\},\gamma\vdash mn$, form a basis for the vector space of all these symmetric functions over $\mathbb{Q}$.

This sum of symmetric functions is denoted by $\{\alpha\}\odot\{\beta\}$:

6.19 $\quad \forall\ \alpha\vdash m,\beta\vdash n\ (\{\alpha\}\odot\{\beta\} := \mathrm{Cyc}(S_n[S_m],(\alpha;\beta)\,|\,\sigma_{1,mn})).$

In fact it is the so-called <u>outer plethysm</u> which Littlewood introduced (Littlewood [5]) in terms of the theory of invariant matrices. Let us call $\{\alpha\} \odot \{\beta\}$ <u>the symmetrized outer product of</u> $\{\alpha\}$ <u>and</u> $\{\beta\}$.

A special case of 6.19 is

6.20 $\quad \{m\} \odot \{n\} = \mathrm{Cyc}(S_n[S_m] \mid \sigma_{1,mn}) = \mathrm{Grf}(S_n[S_m], mn).$

This equation is basic for the application of S-functions in enumeration theory (cf. Read [1]-[5] Foulkes [1]-[3]). (It should be mentioned that conversely the enumeration theory can be applied in order to obtain results on S-functions (cf. P.A. Morris [1]).)

Important applications are the enumeration problems concerning the superposition of graphs (already Redfield's paper deals with this subject). An example will be given in the next section.

Having evaluated the cycle-index polynomial for several permutation groups one may ask how much of the group structure is reflected by the cycle-index polynomial. It does not reflect the complete structure, for nonisomorphic permutation groups may have the same cycle-index. Pólya mentions, that for each odd prime number p and each natural number $m > 2$ there is a non-abelian group of order p^m, each nonidentity element of which is of order p. Hence its regular representation has the cycle-index

$$\frac{1}{p^m}\left(x_1^{p^m} + (p^m-1)x_p^{p^{m-1}}\right).$$

The regular representation of the abelian group $C_p \times \ldots \times C_p$ (m factors, $C_p := \langle(1\ldots p)\rangle$) obviously has the same cycle-index.

Hence the cycle-index reflects only a part of the group structure. In a sense it reflects only the average cycle-structure of the elements.

In this connection one may put the question what can be said about the behaviour of the cycle-structure of a group element $g \in G$ under various permutation representations of G. Golomb and Hales have considered this problem (Golomb/Hales [1]), let us consider their results. They started with the following definitions:

6.21 Def.: Let G be a group and g_1, g_2 be two elements of G. Then g_1 and g_2 are called strongly enumeratively equivalent if an only if their images have the same set of fixed points for every permutation representation of G, and they are called enumeratively equivalent if the numbers of fixed points of $\tau(g_1)$ and $\tau(g_2)$ are equal for all permutation representations τ of G.

Their first results show how these two concepts can be characterized from the group-theoretical point of view:

6.22 If G is a finite group and $g_1, g_2 \in G$, then

(i) g_1 and g_2 are strongly enumeratively equivalent

<u>if and only if they generate the same cyclic subgroup of G, and they are</u>

(ii) <u>enumeratively equivalent if and only if they generate conjugate cyclic subgroups of G.</u>

<u>Proof:</u>

(i) a) Suppose $\langle g_1\rangle = \langle g_2\rangle$ and that $\tau: G \to S_X$ is a permutation representation of G. If $x \in X$ is fixed under $\tau(g_1)$, then $x = \tau(g_1)^{-1}(\tau(g_1)(x)) = \tau(g_1^{-1})(x)$. Hence x is fixed under each power of $\tau(g_1)$, in particular under $\tau(g_2)$.

b) If $\langle g_1\rangle \neq \langle g_2\rangle$, say $g_2 \notin \langle g_1\rangle$, we consider the permutation representation $\tau := I\langle g_1\rangle \uparrow G$, i.e. the permuation representation of G on the left cosets of $\langle g_1\rangle$. The symbol $\langle g_1\rangle$ is fixed under $\tau(g_1)$, but it is not fixed under $\tau(g_2)$.

(ii) a) If $\langle g_2\rangle = h\langle g_1\rangle h^{-1}$, say $g_2 = hg_1^th^{-1}$, and if $x \in X$ is fixed under g_1, then $h(x)$ is fixed under g_2, so that $a_1(g_1) \leq a_1(g_2)$, since h is a bijection. We similarly obtain $a_1(g_2) \leq a_1(g_1)$.

b) If $a_1(\tau(g_1)) = a_1(\tau(g_2))$, for each permutation representation τ of G, we consider $\tau := I\langle g_1\rangle \uparrow G$. Since $\langle g_1\rangle$ is fixed under $\tau(g_1)$, there is (since $a_1(\tau(g_1))=a_1(\tau(g_2))$ a point fixed under $\tau(g_2)$, say the point $h\langle g_1\rangle$. It satisfies $h^{-1}g_2h \in \langle g_1\rangle$, so that $\langle g_2\rangle \subseteq h\langle g_1\rangle h^{-1}$ Analogously we obtain $\langle g_1\rangle \subseteq h'\langle g_2\rangle h'^{-1}$, so that the finiteness of the order $|G|$ of G yields the contradiction:

$$\langle g_2\rangle = h\langle g_1\rangle h^{-1} \text{ for a suitable } h \in G.$$

q.e.d.

Since two group elements generate the same cyclic subgroup if and only if each is a power of the other, Golomb and Hales call such elements relatives, the results 6.22 raise the questions in which cases any two relatives are conjugates and in which cases any two conjugates are relatives. Golomb and Hales obtained part (i) of the following theorem and mentioned that B. Fein had pointed out, that (ii) holds:

6.23 (i) If G is a group, then any two conjugates in G are relatives if and only if each subgroup of G is normal.

(ii) If G is a finite group, then any two relatives are conjugates if and only if all the ordinary characters of G are rational-valued.

Proof:

(i) a) Assume that conjugates are relatives and that $h \in H \leq G \ni g$. Then ghg^{-1} is conjugate to h and hence a relative of h, so that $\langle ghg^{-1}\rangle = \langle h\rangle \leq H$. This implies $H \trianglelefteq G$.

b) If G is a group and $g_1, g_2 \in G$ are conjugates, then since $\langle g_1\rangle$ is normal: $\langle g_1\rangle = \langle g_2\rangle$ so that g_1 and g_2 are relatives.

(ii) There is a theorem (cf. Serre [1], 12.5) saying that all the characters of a finite group G are rational-valued if and only if for each $g \in G$ and $t \in \mathbb{Z}$ with $(t, |\langle g\rangle|) = 1$ g and g^t are conjugates. Hence if relatives are conjugates, each ordinary character has rational values only and vice versa. Golomb and Hales called a group G to be <u>of class</u> $\mathfrak{B}$ if and only if any two relatives in G are conjugates. They obtained the necessary conditions

<u>6.24</u> <u>If $G \in \mathfrak{B}$, then</u>

(i) <u>$1 < |G| < \infty \Rightarrow 2 \mid |G|$, and $G/[G,G]$ has exponent 2,</u>

(ii) <u>$Z(G)$ has exponent 2.</u>

<u>Proof:</u>

(i) If $g \in G\backslash\{1_G\}$, then g and g^{-1} are relatives and hence conjugates since $G \in \mathfrak{B}$. Then, if $g = g^{-1}$, G contains the subgroup $\{1_G, g\}$, so that, since $|G|$ is finite, $|G|$ is even. And if $g \neq g^{-1}$, there is an $h \in G$ which satisfies $g = hg^{-1}h^{-1}$, so that "conjugation by h" is an inner automorphism of even order. Hence $|G|$ is even in both cases.

Furthermore $[g,h] = ghg^{-1}h^{-1} = g^2$ so that for each $g \in G$ g^2 is in $[G,G]$.

(ii) If $g \in Z(G)$ then all its conjugates and hence all its relatives are equal to g itself. Hence $g = g^{-1}$, i.e. $g^2 = 1$.

q.e.d.

Golomb and Hales mention that the symmetric group and the quaternion group of order 8 are of class $\mathcal{B}$, and that together with G and H, the groups $G\times H$ and $G\wr S_2$ are of class $\mathcal{B}$. In the light of the preceding results on characters of wreath products and theorem 6.23 (ii) we complete these results by the following theorem:

<u>6.25</u> <u>Each of the following conditions is sufficient for $G \in \mathcal{B}$:</u>

(i) G is a symmetric group,

(ii) G is the quaternion group of order 8,

(iii) $G \cong H \times I$, H and $I \in \mathcal{B}$,

(iv) $G \cong H\wr I$, where H, $I \in \mathcal{B}$ and I is a permutation group of finite degree n such that $I \cap S_\alpha \in \mathcal{B}$, for all $\alpha \vdash n$.

Since 6.23 (ii) holds we need only mention that the quaternion group of order 8 has the same character table as $S_2\wr S_2$.

7. The construction of patterns

The preceding sections on enumeration under group action were devoted to

(i) the enumeration of the number of patterns which is given by an application of Burnside's lemma to the permutation character (enumeration theorem, constant form),

(ii) the evaluation of cycle-indices which yield (by Pólya-insertion of store enumerator) a generating function for the problem with symmetry group E^H,

(iii) the enumeration theorem in weighted form which enumerates functions $\varphi \in M^N$ by weight with respect to E^H, G^H, $[G]^H$.

We should not leave this subject without saying a word on how these patterns of functions are not only enumerated but even constructed. I.e. we would like to know how to construct a representative for each orbit of the symmetry group in question.

There is in fact a method available which can be used at least for the enumeration problems concerning patterns of functions $\varphi \in M^N$ with respect to E^H, $H \leq S_N$ (see Ruch/Hasselbarth/Richter [1], Brown/Hjelmeland/Masinter [1], Brown/Masinter [1]).

In the case when E^H, for a given $H \leq S_n$, is the symmetry group we can use first the enumeration theorem 5.15 in the weighted form, in order to obtain the number of patterns of functions $\varphi \in M^N$, the elements of which have say r_1 values equal to 1 ($\in M$), r_2 values equal to 2 ($\in M$), ..., r_m values equal to m ($\in M$), so that

$$r_i \in \mathbb{Z}_{\geq 0}, \sum_i r_i = n.$$

Let us assume that there are in fact patterns of this type. We ask for the exact number of patterns of this special type and we would like to construct a representative of each of them.

Instead of the mappings $\varphi \in M^N$ of this type we may consider corresponding elements $\bar{\varphi} \in S_N = S_n$, where $\bar{\varphi}$ is defined as follows. If $i_1 \leq \ldots \leq i_{r_1}$ are the elements of $N = \{1,\ldots,n\}$ which are mapped onto $1 \in M = \{1,\ldots,m\}$, then we put

$$\bar{\varphi}\,(i_\nu) := \nu\ ,\ 1 \leq \nu \leq r_1.$$

If $j_1 \leq \ldots \leq j_{r_2}$ are the elements of N which are mapped onto $2 \in M$, then we put

$$\bar{\varphi}\,(j_\nu) := r_1 + \nu,\ 1 \leq \nu \leq r_2.$$

In terms of these elements $\bar{\varphi} \in S_n$ the equivalence of φ, ψ reads as follows (recall that both φ and ψ have r_i values equal i, $1 \leq i \leq m$): φ and $\psi \in M^N$ are elements of the same pattern (with type $(r_1,\ldots,r_m)$) if and only if there is a $\sigma \in S_{r_1} \times \ldots \times S_{r_m}$ (or $S_{r_1} \oplus \ldots \oplus S_{r_m}$ if you like) and a $\pi \in H$ which satisfy

$$\bar{\psi} = \sigma \circ \bar{\varphi} \circ \pi^{-1}.$$

In other words:

7.1 <u>φ and $\Psi \in M^N$ of type $(r_1,\dots,r_m)$ are representatives of the same pattern if and only if $\bar{\varphi}$ and $\bar{\Psi}$ belong to the same double coset</u>

$$\underline{S_{r_1}\times \dots \times S_{r_m}\ \gamma\ H}$$

<u>of $S_{r_1} \times \dots \times S_{r_m}$ and H in S_n.</u>

Let us illustrate that by the introductory necklace problem. We ask for a construction of all the necklace with five beads three of which are of colour • and the remaining two of colour ∘.

Evaluation of $Cyc(D_5|x+y)$ yields 5.1, where 2 is the coefficient of x^3y^2 so that there in fact two necklaces of type $(r_1,r_2) = (3,2)$ exist. We are left with the question how they can be constructed.

The preceding considerations have shown that a complete system $\bar{\varphi}_1, \dots, \bar{\varphi}_s$ of representatives of the double cosets $S_{r_1} \times \dots \times S_{r_n}\ \gamma\ H$ yield a complete system $\varphi_1, \dots, \varphi_s$ of the patterns of type $(r_1,\dots,r_m)$. In case of the necklace problem we have to evaluate a system of representations of the double cosets of

$$S_3 \times S_2 = \{1,(12),(13),(23),(123),(132)\} \times \{1,(45)\}$$

and

$$D_5 = \{1, (12345),(13524),(14253),(15432),$$
$$(25)(34), (13)(45), (24)(15), (12)(35), (14)(23)\}$$

in S_5.

It turns out that $\{1,(14532)\}$ is such a system, i.e. we obtain

$$\bar{\varphi}_1 = \begin{pmatrix} 1 & 2 & 3 & 4 & 5 \\ 1 & 2 & 3 & 4 & 5 \end{pmatrix}$$

and

$$\bar{\varphi}_2 = \begin{pmatrix} 1 & 2 & 3 & 4 & 5 \\ 4 & 1 & 2 & 5 & 3 \end{pmatrix}$$

The corresponding mappings are

$$\varphi_1: \quad \begin{aligned} 1 &\mapsto \bullet \\ 2 &\mapsto \bullet \\ 3 &\mapsto \bullet \\ 4 &\mapsto \circ \\ 5 &\mapsto \circ \end{aligned} \qquad\qquad \varphi_2: \quad \begin{aligned} 1 &\mapsto \circ \\ 2 &\mapsto \bullet \\ 3 &\mapsto \bullet \\ 4 &\mapsto \circ \\ 5 &\mapsto \bullet \end{aligned}$$

The necklaces are

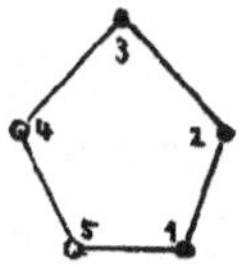

and

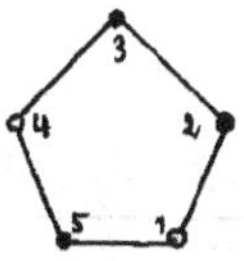

It is not very difficult to make a computer produce such a complete system of representatives of double cosets if the degree n of the group H is relatively small. One may use for this a Todd-Coxeter algorithm which produces a complete system of representatives of left cosets of H first.

Tricky methods which can be applied for higher degrees can be found in the papers Brown/Hjelmeland/Masinter [1] and Brown/Masinter [1] where also special applications to chemical structure elucidation are discussed.

Let us return to 7.1. It shows that the number of pattern of type $(r_1,\dots,r_m)$ is equal to the number of double cosets $S_{r_1} \times \dots \times S_{r_m} \gamma H$ in S_n. The preceding sections have shown that this number of patterns is also equal to the coefficient of $x_1^{r_1} \dots x_m^{r_m}$ in $Cyc(H|\sigma_{1,m}) = Grf(H,m)$. This yields:

7.2 <u>The number of double cosets $S_{r_1} \times \dots \times S_{r_m} \gamma H$ in S_n is the coefficient of $x_1^{r_1} \dots x_m^{r_m}$ in $Grf(H,m)$.</u>

<u>In other words: $Grf(H,m) = Cyc(H|\sigma_{1,m})$ is the generating function for double cosets $S_{r_1} \times \dots \times S_{r_m} \gamma H$ in S_n.</u>

Recall that 5.22 yields

7.3 $$Cyc(H|\sigma_{1,m}) = Cyc(S_n, IH \uparrow S_n \,|\sigma_{1,m})$$

$$= \sum_{\alpha \vdash n} (IH \uparrow S_n, [\alpha]) \cdot \{\alpha\}.$$

A generalization of this is discussed in Littlewood's book on group characters in chapter IX ("Structure of groups"), section 5 ("Transitivity").

If both G and H denote subgroups of S_n, then the number of double cosets $G \pi H$ in S_n is called <u>the transitive factor of</u> G <u>and</u> H and denoted by

$$N(G,H).$$

Littlewood then shows that the following holds:

7.4 $$N(G,H) = \sum_{\alpha \vdash n} (IG \uparrow S_n, [\alpha]) \cdot (IH \uparrow S_n, [\alpha]).$$

This result has interesting corollaries concerning group reduction functions of multiply transitive groups. Before we discuss these corollaries let us consider, how N(G,H) fits into the preceding considerations.

This consideration leads to Redfield's paper (Redfield [1], cf. also Foulkes [1], [2], Harary/Palmer [3]). It is devoted to the enumeration of what he calls group-reduced distributions.

The starting point is again the set W of $n \times m$ - matrices $A = (a_{ik})$, each row of which contains the m elements of $M := \{1,\ldots,m\}$ (cf. section 6). Hence

$$|W| = m!^n.$$

Redfield calls the set M the <u>range.</u>

Every column of an element $A \in W$ constitutes a correspondence between n elements each of which belongs to the range M. Hence we call two elements of W equivalent if they constitute the same n-tuple of correspondences, we define

7.5 $(a_{ik}) \sim (b_{ik}) \quad :\Leftrightarrow \exists\, \pi \in S_m \; ((a_{ik}) = (b_{i,\pi^{-1}(k)}))$.

I.e. $A \sim B$ if and only if A arises from B by a permutation of columns. We notice that

$$|W/\sim| = m!^{n-1}.$$

Such a class, an element of $W/\sim$, is called a <u>range-correspondence.</u>

If we are given subgroups $G_1, \ldots, G_n$ of S_m, then they define an equivalence relation on $W/\sim$ as follows:

7.6 $[A]_\sim \approx [B]_\sim \quad :\Leftrightarrow \forall\, 1 \leq i \leq n \;\exists \rho_i \in G_i \; ([(a_{ik})]_\sim = [(\rho_i b_{ik})]_\sim)$.

We ask for the number

$$|(W/\sim)/\approx|$$

of equivalence classes. The classes are called <u>group-reduced distributions.</u> Considering the special element

$$A_o := \begin{pmatrix} 1 & \dots & m \\ \dots & \dots & \dots \\ 1 & \dots & m \end{pmatrix} \in W,$$

we see that for each $A \in W$ there are uniquely determined $\pi_i^A \in S_n$ which satisfy

$$A = (\pi_i^A(k)) =: (\pi_1^A, \dots, \pi_n^A) A_o.$$

7.6 then reads as follows:

7.7 $[A]_\sim \approx [B]_\sim \Leftrightarrow \exists\, \rho_i \in G_i,\ \sigma \in S_m((\pi_1^A, \dots, \pi_n^A) = (\rho_i \pi_1^B \sigma, \dots, \rho_n \pi_n^B \sigma))$

Hence

$$[A]_\sim \approx [B]_\sim \Leftrightarrow \exists\, \sigma \in S_m\, ((\pi_1^A, \dots, \pi_n^A) \in G_1 \pi_1^B \sigma \times \dots \times G_n \pi_n^B \sigma).$$

This shows that Redfields group-reduced distributions are in a sense the orbits of the permutation group induced by S_m on the cartesian product

$$\mathop{\times}_i S_m/G_i$$

of the sets S_m/G_i of right cosets of the G_i in S_m. The permutation group has in fact the character

$$\bigotimes_{i=1}^{n} (I\, G_i \uparrow S_m),$$

so that in terms of representation theory we obtain forthe

desired number of group-reduced distributions:

7.8 $\left|{}^{(W/\sim)}/_{\approx}\right| = \left(\bigotimes_i (IG_i \uparrow S_m), [m]\right).$

If in particular n := 2, then this number is equal to

$$\left(\bigotimes_1^2 (IG_i \uparrow S_m), [m]\right) = (IG_1 \uparrow S_m, IG_2 \uparrow S_m),$$

which is the number of double cosets $G_1 \pi G_2$ in S_m (use Mackey's theorem or prove it by noticing that an element of W/~ is in a sense just a permutation of S_m if n = 2). Hence 7.8 yields 7.4 as a corollary. This can be applied to multiply transitive groups G_1.

We use the following lemma:

7.9 $G \leq S_m$ is k-fold transitive if and only if $S_{m-k} \cdot G = S_m$, i.e. if and only if $N(S_{m-k}, G) = 1$.

7.8 yields now as a corollary:

7.10 $G \leq S_m$ is k-fold transitive if and only if $1 = ([m-k] \uparrow S_m, IG \uparrow S_m)$.

The branching rule (I 4.52) and 7.10 imply (since $IG \uparrow S_m$ contains the identity representation [m] exactly once as does $[m-k] \uparrow S_m$):

7.11 $G \leq S_m$ is k-fold transitive if and only if

$\forall\, \alpha \vdash m\ (m-k \leq \alpha_1 < m \Rightarrow (IG\uparrow S_m, [\alpha]) = 0)$.

In terms of the group reduction function (apply 5.17, 5.23) we obtain (Littlewood [2]):

7.12 $G \leq S_m$ is k-fold transitive if and only if

$\forall\, \alpha \vdash m\ (m-k \leq \alpha_1 < m \Rightarrow \{\alpha\}$ is not a summand of $\mathrm{Grf}(G))$.

Since $\mathrm{Grf}(S_m) = \{m\}$ and $\mathrm{Grf}\,(A_m) = \{m\} + \{1^m\}$, this also yields that S_m is m-fold transitive and that A_m is (m-2)-fold transitive but not (m-1)-fold transitive. 7.12 can be used in order to evaluate group reduction functions.

Let us return to 7.8 and discuss some of its applications. 7.8 can be used in order to solve problems concerning the superposition of graphs. Redfield observed that the number of different superpositions of graphs $\Gamma_1, \ldots, \Gamma_n$ with m points and with automorphismen groups $G_1, \ldots, G_n$ is just 7.8.

If for example

$\Gamma_1 := \Gamma_2 := \Gamma_3 :=$ 

so that

$$G_1 = G_2 = G_3 = S_2[S_2]$$

we obtain for the number of superpositions:

$$\left(\bigotimes_{1}^{3}(IG_i \uparrow S_4), [4]\right) = \left(\bigotimes^{3}((2;2) \uparrow S_4), [4]\right)$$

$$= \left(\bigotimes^{3}([2] \odot [2]), [4]\right) = \left(\bigotimes^{3}([4] + [2^2]), [4]\right) = 5$$

(use the character table of S_4 and the decomposition of $[2]\odot[2]$).
These 5 superpositions are indicated below:

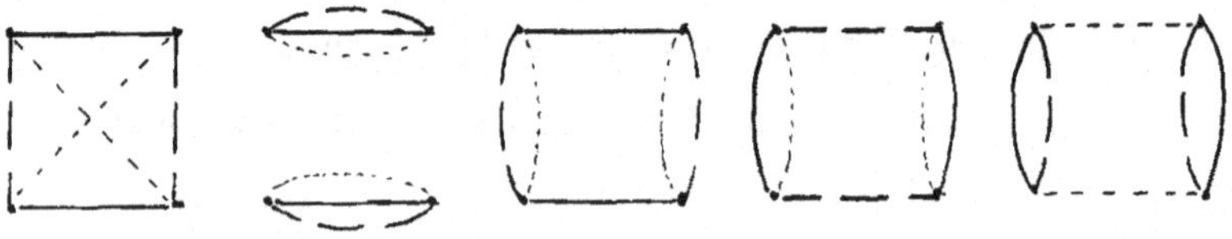

Fig. 4

References (in addition to the references in Vol. I)

Benson, C.T./ Curtis, C.W.: [1] On the degrees and rationality of certain characters of finite Chevalley groups. Trans. Amer. Math. Soc. 165 (1972), 251-273.

Berge, C.: [1] Principles of combinatorics. Academic Press, New York, 1971.

Brown, H./ Hjelmeland, L./ Masinter, L.: [1] Constructive graph labeling using double cosets Discrete Math. 7 (1974), 1-30.

Brown, H./ Masinter, L.: [1] An Algorithm for the construction of the graphs of organic molecules. Discrete Math. 8 (1974), 227-244.

deBruijn, N.G.: [1] Pólya's theory of counting. Applied Combinatorial Mathematics (E.F. Beckenbach, ed.), 144-184. Wiley, New York, 1964.

[2] Generalization of Pólya's fundamental theorem in enumerative combinatorial analysis. Nederl. Akad. Wetensch. Proc. Ser. A 62 (1959), 59-69.

[3] The exterior cycle index of a permutation group. Studies in Pure Math. (Presented to Richard Rado). 31-37. Academic Press 1971.

[4] A survey of generalizations of Pólya's enumeration theorem. Nieuw Arch. Wiskunde 19 (1970), 89-112.

[5] Pólya's Abzähl-Theorie: Muster für Graphen und chemische Verbindungen. Selecta Math., ed. K. Jacobs, vol. III, 1-26. Springer-Verlag 1971.

Butler, P.H.: [1] Coupling coefficients and tensor operators for chains of groups. Trans. Roy. Soc. London (to appear)

Butler, P.H./King, R.C.: [1] Branching rules for U(N)⊃U(M) and the evaluation of outer plethysms. J. Math. Phys. 14 (1973), 741-745.

[2] The symmetric group: Characters, products and plethysms. J. Math. Phys. 14 (1973), 1176-1183.

[3] Symmetrized Kronecker products of representations. Can. J. Math. 26 (1974), 328-339.

Carter, R.W.: [1] Conjugacy classes in the Weyl group. Seminar on Algebraic Groups and Related Finite Groups. The Institute for Advanced Study, Princeton, N.J., 1968/69, pp. 297-318. Lecture Notes in Math. vol. 131, Springer, Berlin 1970.

[2] Conjugacy classes in the Weyl group. Compositio Math. 25 (1972), 1-59.

Cayley, A.: [1] On the mathematical theory of isomers. Philos. Magazine 47 (1874), 444-446.

[2] On the analytical forms called trees, with application to the theory of chemical combinations Report of British Association for the Advancement of Science, 1875, 257-305

Celik, Ö./ Kerber, A./ Pahlings, H.: [1] Zur Darstellungstheorie gewisser Verallgemeinerungen der Serien von Weyl-Gruppen (in preparation)

Clausen, M. [1] Zentralisatorenverbände von Moduln über halbeinfachen Gruppenalgebren; zur Theorie der Symmetrieklassen in Tensorräumen. Diplomarbeit, Gießen 1974.

Comtet, L.: [1] Advanced Combinatorics. D. Reidel Publishing Company 1974.

Derome, J.-R.: [1] Symmetry properties of the 3j-symbols for an arbitrary group. J. Math. Phys. 7 (1966), 612-61

Dieudonné, J.A./ Carrell, J.B.: [1] Invariant theory, old and new. Academic Press, New York, 1971.

Esper, N.: [1] Ein interaktives Programmsystem zur Erzeugung der rationalisierten Charakterentafel einer endlichen Gruppe. Staatsexamensarbeit, Aachen 1974.

[2] Tables of reductions of symmetrized inner products ("inner plethysms") of ordinary irreducible representations of symmetric groups. (to appear)

Feit, W.: [1] Characters of finite groups. W.A. Benjamin Inc., 1967.

Foulkes, H.O.: [1] On Redfield's group reduction functions. Canad. Math. 15 (1963), 272-284.

[2] On Redfield's range-correspondences. Canad. J. Math. 18 (1966), 1060-1071.

[3] Linear graphs and Schur-functions. Conference on Comb. Math., Oxford 1969.

Frobenius, F.G.: [1] Über die Charaktere der symmetrischen Gruppe. Sitzgsber. Preuß. Akad. Wiss. 1900, 516-534.

Golomb, S.W./ Hales, A.W.: [1] On Enumerative Equivalence of Group Elements. J. Comb. Theory 5 (1968), 308-312

Gretschel, B.: [1] Berechnung der Charakterentafeln von Symmetrien symmetrischer Gruppen. Diplomarbeit Gießen, 1973.

Harary, F./ Beineke, L.: [1] A seminar on graph theory. Holt, Rinehart and Winston, New York, 1967

Harary, F./ Palmer, E.: [1] Graphical enumeration. Academic Press, New York, 1973.

[2] The power group enumeration theorem. J. Comb. Theory 1 (1966), 157-173.

[3] The enumeration methods of Redfield. Amer. J. Math. 89 (1967), 373-384.

Hilge, A.: [1] Berechnung der Charakterentafeln von Symmetrien symmetrischer Gruppen. Diplomarbeit Gießen, 1973.

Humphreys, J.E.: [1] Introduction to Lie Algebras and Representation Theory. Graduate Texts in Mathematics 9, Springer-Verlag, Berlin 1972, xii + 169 pp.

Kerber,A.: [8] Characters of wreath products and applications to representation theory and combinatorics. (to appear in 'Discrete Mathematics')

[9] Der Zykelindex der Exponentialgruppe. Mitt. math. Sem. Univ. Gießen 98 (1973), 5-20.

Klaiber,B.: [1] Fortsetzbarkeit und Korrespondenz von Darstellungen. Dissertation, Mainz 1969.

King,R.C.: [1] Branching rules for GL(N) $\supset \Sigma_m$ and the evaluation of inner plethysms. J. Math. Phys. 15 (1974), 258-267.

Knutson, D.: [1] λ-Rings and the representation theory of the symmetric group. Lecture Notes in Math., vol. 308 Springer-Verlag 1973.

Kostant,B.: [1] A theorem of Frobenius, a theorem of Amitsur-Levitski and cohomology theory. J. Math. Mech. 7 (1958), 237-264

Lehmann,W.: [1] Das Abzählltheorem der Exponentialgruppe in gewichteter Form. Mitt.math.Sem.Univ. Gießen. 112

Littlewood,D.E.: [4] Plethysm and the inner product of S-functions. J. London Math. Soc. 32 (1957), 18-22.

[5] Polynomial concomitants and invariant matrices. J.London Math. Soc. 11 (1936), 49-55

Livingstone, D./Wagner, A.: [1] Transitivity of finite permutation groups on unordered sets. Math. Z. 90 (1965), 393-405.

Liu,C.L.: [1] Introduction to combinatorial mathematics. McGraw-Hill, New York, 1968

Mayer,S.J.: [1] On the irreducible characters of the Weyl groups. Ph.D. thesis, University of Warwick, 1971.

Morris,P.A.: [1] Applications of graph theory to S-function theory. J. London Math.Soc. 8 (1974), 63-72

Murnaghan, F.D.: [1] On the representations of the symmetric group. Amer. J. Math. 59 (1937), 437-488.

[2] The characters of the symmetric group. Amer. J. Math. 59 (1937), 739-753.

[3] The analysis of the Kronecker product of irreducible representations of the symmetric group. Amer. J. Math. 60 (1938), 761-784.

Palmer, E./ Robinson, R.W.: [1] Enumeration under two representations of the wreath product. Acta Math. 131 (1973), 123-143.

[2] The matrix group of two permutation groups. Bull. Amer. Math. Soc. 73 (1967), 204-207

Read, R.C.: [1] The enumeration of locally restricted graphs I. J. London Math. Soc. 34 (1959), 417-436

[2] The enumeration of locally restricted graphs II. J. London Math. Soc. 35 (1960), 344-351.

[3] On the number of self-complementary graphs and digraphs. J. London Math. Soc. 38 (1963), 99-104.

[4] Some applications of a theorem of deBruijn. Graph Theory and Theoretical Physics (ed. F. Harary), 273-280. Academic Press 1967.

[5] The use of S-functions in combinatorical analysis. Canad. J. Math. 20 (1968), 808-841.

Redfield, J.H.: [1] The theory of group-reduced distributions. Amer. J. Math. 49 (1927), 433-455.

Reynolds, W.F.: [1] Fields related to Brauer characters. Math. Z. 135 (1974), 363-367.

Robinson, G. de B.: [7] On a theorem of Livingstone and Wagner. Math. Z. 102 (1967), 351-352.

Ruch, E./ Hasselbarth, W./ Richter, B.: [1] Doppelnebenklassen als Klassenbegriff und Nomenklaturprinzip für Isomere und ihre Abzählung. Theor. Chim. Acta 19 (1970), 288-300.

Rudvalis, A./ Snapper, E.: [1] Permutation representations of finite groups. Mimeographed notes.

Sänger, F.: [1] Einige Charakterentafeln von Symmetrien symmetrischer Gruppen. Mitt. math. Sem. Univ. Gießen 98 (1973), 21-38

Schur, I.: [1] Über eine Klasse von Matrizen, die sich einer gegebenen Matrix zuordnen lassen. Dissertation, Berlin 1901.

[2] Über die rationalen Darstellungen der allgemeinen linearen Gruppe. Sitzgsber. preuß. Akad. Wiss. 1927, 58-75.

Serre, J.-P.: [1] Représentations linéaires des groupes finis. Hermann, Paris, 2e edition 1967, 182 pp.

Stewart, J.: [1] Lie Algebras. Lecture Notes in Mathematics 127, Springer-Verlag, Berlin, 1970, pp. 97

Taylor, T.: [1] Representation theory of the symmetric and hyper-octohedral groups. Dissertation, Aberystwyth.

[2] Representations of Weyl groups. Thesis, Aberystwyth 1973.

van der Waerden, B.: [1] Der Zusammenhang zwischen den Darstellungen der symmetrischen und der linearen Gruppen. Math. Ann. 104 (1931), 92-95, 800.

Weyl, H.: [1] Theorie der Darstellung kontinuierlicher halbeinfacher Gruppen durch lineare Transformationen. Math. Z. 23 (1925), 271-309.

[2] Der Zusammenhang zwischen der symmetrischen und der linearen Gruppe. Ann. of Math. 30 (1929), 449-516.

[3] Commutator algebra of a finite group of collineations. Duke Math. J. 3 (1937), 200-212.

Wille, D.: [1] On the enumeration of self-complementary m-placed relations. Discrete Math. 10 (1974), 189-192.

Williamson, S.G.: [1] Symmetry Operators of Kranz Products. J. Comb. Theory 11 (1971), 122-138.

[2] Operator Theoretic Invariants and the Enumeration Theory of Pólya and de Bruijn. J. Comb. Theory 8 (1970), 162-169.

van Zanten, A.J./[1] On the number of roots of the equation $x^n = 1$
de Vries, E.: in finite groups and related properties. J. Algebra 25 (1973), 475-486.

Subject Index

Vol. 342: Algebraic K-Theory II, "Classical" Algebraic K-Theory, and Connections with Arithmetic. Edited by H. Bass. XV, 527 pages. 1973.

Vol. 343: Algebraic K-Theory III, Hermitian K-Theory and Geometric Applications. Edited by H. Bass. XV, 572 pages. 1973.

Vol. 344: A. S. Troelstra (Editor), Metamathematical Investigation of Intuitionistic Arithmetic and Analysis. XVII, 485 pages. 1973.

Vol. 345: Proceedings of a Conference on Operator Theory. Edited by P. A. Fillmore. VI, 228 pages. 1973.

Vol. 346: Fučík et al., Spectral Analysis of Nonlinear Operators. II, 287 pages. 1973.

Vol. 347: J. M. Boardman and R. M. Vogt, Homotopy Invariant Algebraic Structures on Topological Spaces. X, 257 pages. 1973.

Vol. 348: A. M. Mathai and R. K. Saxena, Generalized Hypergeometric Functions with Applications in Statistics and Physical Sciences. VII, 314 pages. 1973.

Vol. 349: Modular Functions of One Variable II. Edited by W. Kuyk and P. Deligne. V, 598 pages. 1973.

Vol. 350: Modular Functions of One Variable III. Edited by W. Kuyk and J.-P. Serre. V, 350 pages. 1973.

Vol. 351: H. Tachikawa, Quasi-Frobenius Rings and Generalizations. XI, 172 pages. 1973.

Vol. 352: J. D. Fay, Theta Functions on Riemann Surfaces. V, 137 pages. 1973.

Vol. 353: Proceedings of the Conference on Orders, Group Rings and Related Topics. Organized by J. S. Hsia, M. L. Madan and T. G. Ralley. X, 224 pages. 1973.

Vol. 354: K. J. Devlin, Aspects of Constructibility. XII, 240 pages. 1973.

Vol. 355: M. Sion, A Theory of Semigroup Valued Measures. V, 140 pages. 1973.

Vol. 356: W. L. J. van der Kallen, Infinitesimally Central Extensions of Chevalley Groups. VII, 147 pages. 1973.

Vol. 357: W. Borho, P. Gabriel und R. Rentschler, Primideale in Einhüllenden auflösbarer Lie-Algebren. V, 182 Seiten. 1973.

Vol. 358: F. L. Williams, Tensor Products of Principal Series Representations. VI, 132 pages. 1973.

Vol. 359: U. Stammbach, Homology in Group Theory. VIII, 183 pages. 1973.

Vol. 360: W. J. Padgett and R. L. Taylor, Laws of Large Numbers for Normed Linear Spaces and Certain Fréchet Spaces. VI, 111 pages. 1973.

Vol. 361: J. W. Schutz, Foundations of Special Relativity: Kinematic Axioms for Minkowski Space-Time. XX, 314 pages. 1973.

Vol. 362: Proceedings of the Conference on Numerical Solution of Ordinary Differential Equations. Edited by D.G. Bettis. VIII, 490 pages. 1974.

Vol. 363: Conference on the Numerical Solution of Differential Equations. Edited by G. A. Watson. IX, 221 pages. 1974.

Vol. 364: Proceedings on Infinite Dimensional Holomorphy. Edited by T. L. Hayden and T. J. Suffridge. VII, 212 pages. 1974.

Vol. 365: R. P. Gilbert, Constructive Methods for Elliptic Equations. VII, 397 pages. 1974.

Vol. 366: R. Steinberg, Conjugacy Classes in Algebraic Groups (Notes by V. V. Deodhar). VI, 159 pages. 1974.

Vol. 367: K. Langmann und W. Lütkebohmert, Cousinverteilungen und Fortsetzungssätze. VI, 151 Seiten. 1974.

Vol. 368: R. J. Milgram, Unstable Homotopy from the Stable Point of View. V, 109 pages. 1974.

Vol. 369: Victoria Symposium on Nonstandard Analysis. Edited by A. Hurd and P. Loeb. XVIII, 339 pages. 1974.

Vol. 370: B. Mazur and W. Messing, Universal Extensions and One Dimensional Crystalline Cohomology. VII, 134 pages. 1974.

Vol. 371: V. Poenaru, Analyse Différentielle. V, 228 pages. 1974.

Vol. 372: Proceedings of the Second International Conference on the Theory of Groups 1973. Edited by M. F. Newman. VII, 740 pages. 1974.

Vol. 373: A. E. R. Woodcock and T. Poston, A Geometrical Study of the Elementary Catastrophes. V, 257 pages. 1974.

Vol. 374: S. Yamamuro, Differential Calculus in Topological Linear Spaces. IV, 179 pages. 1974.

Vol. 375: Topology Conference. Edited by R. F. Dickman Jr. and P. Fletcher. X, 283 pages 1974.

Vol. 376: I. J. Good and D. B. Osteyee, Information, Weight of Evidence. The Singularity between Probability Measures and Signal Detection. XI, 156 pages. 1974.

Vol. 377: A. M. Fink, Almost Periodic Differential Equations. VIII, 336 pages. 1974.

Vol. 378: TOPO 72 – General Topology and its Applications. Proceedings 1972. Edited by R. A. Alò, R. W. Heath and J. Nagata. XIV, 651 pages. 1974.

Vol. 379: A. Badrikian et S. Chevet, Mesures Cylindriques, Espaces de Wiener et Fonctions Aléatoires Gaussiennes. X, 383 pages. 1974.

Vol. 380: M. Petrich, Rings and Semigroups. VIII, 182 pages. 1974.

Vol. 381: Séminaire de Probabilités VIII. Edité par P. A. Meyer. IX, 354 pages. 1974.

Vol. 382: J. H. van Lint, Combinatorial Theory Seminar Eindhoven University of Technology. VI, 131 pages. 1974.

Vol. 383: Séminaire Bourbaki – vol. 1972/73. Exposés 418–435. IV, 334 pages. 1974.

Vol. 384: Functional Analysis and Applications, Proceedings 1972. Edited by L. Nachbin. V, 270 pages. 1974.

Vol. 385: J. Douglas Jr. and T. Dupont, Collocation Methods for Parabolic Equations in a Single Space Variable (Based on C^1-Piecewise-Polynomial Spaces). V, 147 pages. 1974.

Vol. 386: J. Tits, Buildings of Spherical Type and Finite BN-Pairs. X, 299 pages. 1974.

Vol. 387: C. P. Bruter, Eléments de la Théorie des Matroïdes. V, 138 pages. 1974.

Vol. 388: R. L. Lipsman, Group Representations. X, 166 pages. 1974.

Vol. 389: M.-A. Knus et M. Ojanguren, Théorie de la Descente et Algèbres d' Azumaya. IV, 163 pages. 1974.

Vol. 390: P. A. Meyer, P. Priouret et F. Spitzer, Ecole d'Eté de Probabilités de Saint-Flour III – 1973. Edité par A. Badrikian et P.-L. Hennequin. VIII, 189 pages. 1974.

Vol. 391: J. W. Gray, Formal Category Theory: Adjointness for 2-Categories. XII, 282 pages. 1974.

Vol. 392: Géométrie Différentielle, Colloque, Santiago de Compostela, Espagne 1972. Edité par E. Vidal. VI, 225 pages. 1974.

Vol. 393: G. Wassermann, Stability of Unfoldings. IX, 164 pages. 1974.

Vol. 394: W. M. Patterson, 3rd, Iterative Methods for the Solution of a Linear Operator Equation in Hilbert Space – A Survey. III, 183 pages. 1974.

Vol. 395: Numerische Behandlung nichtlinearer Integrodifferential- und Differentialgleichungen. Tagung 1973. Herausgegeben von R. Ansorge und W. Törnig. VII, 313 Seiten. 1974.

Vol. 396: K. H. Hofmann, M. Mislove and A. Stralka, The Pontryagin Duality of Compact O-Dimensional Semilattices and its Applications. XVI, 122 pages. 1974.

Vol. 397: T. Yamada, The Schur Subgroup of the Brauer Group. V, 159 pages. 1974.

Vol. 398: Théories de l'Information, Actes des Rencontres de Marseille-Luminy, 1973. Edité par J. Kampé de Fériet et C.-F. Picard. XII, 201 pages. 1974.